TRIGONOMETRY

Trigonometry

FRANK C. DENNEY
Chabot College

CANFIELD PRESS ⌽ SAN FRANCISCO
A Department of Harper & Row, Publishers, Inc.
New York   Hagerstown   London

Design and chapter-opening art: Paul Quin
Line art: Larry Jansen
Cover: Paul Quin

LIBRARY OF CONGRESS CATALOGING IN PUBLICATION DATA
Denney, Frank C
　Trigonometry.

　Includes index.
　1. Trigonometry, Plane.　　I. Title.
QA533.D38　　516'.24　　75-25692
ISBN 0-06-382560-0

TRIGONOMETRY
Copyright © 1976 by Frank C. Denney
Printed in the United States of America. All rights reserved.
No part of this book may be used or reproduced in any manner
whatsoever without written permission except in the case of
brief quotations embodied in critical articles and reviews.
For information address Harper & Row, Publishers, Inc.,
10 East 53rd St., New York, N.Y. 10022.

76　77　78　10　9　8　7　6　5　4　3　2　1

Contents

CHAPTER ONE

Preparing for Trigonometry

1.1 INTRODUCTION 1
1.2 RIGHT TRIANGLE TRIGONOMETRY 2
1.3 A PREVIEW OF THINGS TO COME 10
1.4 FACTS AND FORMULAS FROM GEOMETRY 14
1.5 FACTS AND FORMULAS FROM ALGEBRA 21
1.6 ANGLE MEASUREMENTS 28
  REVIEW EXERCISES 33

CHAPTER TWO

Trigonometric Functions

2.1 THE TRIGONOMETRIC RATIOS; FUNDAMENTAL IDENTITIES 35
2.2 SPECIAL ANGLES 40
2.3 FUNCTIONS OF ANY ANGLE 46
2.4 TRIGONOMETRIC FUNCTIONS AND THEIR GRAPHS 51
  REVIEW EXERCISES 57

CHAPTER THREE

Identities, Formulas, and Equations

3.1 PROVING IDENTITIES 58
3.2 ADDITION AND SUBTRACTION IDENTITIES 65
3.3 DOUBLE AND HALF-ANGLE FORMULAS 73
3.4 CONDITIONAL EQUATIONS 79
  REVIEW EXERCISES 85

CHAPTER FOUR

Graphs of Trigonometric Functions and Inverse Functions

4.1 GRAPHS OF $y = A \sin Bx$ AND $y = A \cos Bx$ 87
4.2 GRAPHS OF $y = A \sin B(x + C)$ AND $y = A \cos B(x + C)$ 95
4.3 SKETCHING THE NONSINUSOIDAL TRIGONOMETRIC FUNCTIONS   99
4.4 GRAPHS OF RELATIONS, FUNCTIONS, AND THEIR INVERSES   104
4.5 THE INVERSE TRIGONOMETRIC FUNCTIONS   109
4.6 ADDITION OF ORDINATES   114
    REVIEW EXERCISES   119

CHAPTER FIVE

Complex Numbers and Polar Coordinates

5.1 COMPLEX NUMBERS IN ALGEBRAIC FORM   121
5.2 GRAPHICAL REPRESENTATION; TRIGONOMETRIC FORM OF COMPLEX NUMBERS   127
5.3 POWERS AND ROOTS OF COMPLEX NUMBERS   132
5.4 POLAR COORDINATES   138
    REVIEW EXERCISES   144

CHAPTER SIX

Applications of Trigonometry

6.1 THE LAW OF SINES   146
6.2 THE LAW OF SINES: AMBIGUOUS CASE   153
6.3 THE LAW OF COSINES   159
6.4 VECTORS   166
6.5 MISCELLANEOUS APPLICATIONS   173
    REVIEW EXERCISES   179

APPENDIXES
   A INTERPOLATION   180
   B COMPUTING VALUES OF TRIGONOMETRIC FUNCTIONS   186
   C LOGARITHMS OF THE TRIGONOMETRIC FUNCTIONS   188
   D TABLES   I   Trigonometric Functions, Values in Degrees   196
              II   Trigonometric Functions, Values in Radians   204
              III   Logarithms, Base 10   208
              IV   Logarithms of Trigonometric Functions   210
              V   Squares and Square Roots   218

ANSWERS TO SELECTED PROBLEMS   219

INDEX   239

Preface

A friend who is an engineer asked me why I was writing a trigonometry textbook. The subject has been around for quite a long time and there are plenty of books on the subject available, so why should I spend so much time writing about something so basic?

My answer comes from my long experience as a teacher of mathematics. I have used some good texts and some poor ones, and have seldom been completely satisfied with even the best of the good ones. I suspect that every instructor has had the same experience. Students, whose exposure to textbooks is somewhat more limited, have a hard time distinguishing between the difficulties of the subject matter and the difficulties inherent in the text.

My purpose has been to write a book that can be read and understood by students. It is not a programmed text. It will probably be used in traditional courses with an instructor who gives lectures and assigns homework. However, I have prepared the text with examples and explanations in such a way that a student can successfully study the material independently.

Students need to want to learn in order to succeed. Sometimes the motivation comes from outside the subject itself—completion of prerequisites, requirements for a major or minor, maintaining a grade point average—but it is better if students can be interested in the subject itself. To this end I have started the text with an introduction to right triangle trigonometry, a topic so simple and yet so powerful and useful that the reader can immediately see the answer to the question, "Why study trigonometry?"

Designed to be used in a one-quarter or one-semester course in college trigonometry, this text can be studied by students who have had one and one-half years of high school algebra or the equivalent. Geometry would be a helpful prerequisite, but those who are willing to study a little harder can probably manage without it, provided they read the section on geometry in Chapter One very carefully.

Throughout the text I have included applications as often as appropriate, and Chapter Six is concerned completely with applications. Graphs of trigonometric functions seem to be difficult for many, so I have introduced graphing early, at the end of Chapter Two. I have done so in a very elementary way, expecting that the graphs will be plotted point by point by the student, so that he or she will see how they develop and then later appreciate the methods for rapid sketching presented in Chapter Four. Graphs of the inverse functions are covered in more explicit detail here than in most trigonometry texts. Although inverse functions are generally introduced in the prerequisite algebra courses, the application to trigonometric functions is frequently far from obvious to students, so I have included a preparatory section that should make the transition relatively easy.

The exercises are carefully balanced, so that those who do either all of the odd-numbered problems or all of the even-numbered problems will cover the material. However, the even-numbered exercises are a shade more difficult and include most of the required derivations or proofs. The answers for most of the odd-numbered exercises and for all of the chapter review exercises are in the back of the book. The answers to the even-numbered problems can be found in the Instructor's Manual.

Easy access to electronic calculators has diminished the necessity for both interpolation and logarithms of trigonometric functions, but these topics are included in the appendix for those who would like to study them. The tables are standard tables, presented in an easy-to-read format. Although many students today have "scientific" calculators, which give the values of trigonometric functions to eight or more significant digits, the answers for this text were calculated using the degree of accuracy available from the tables of the text. Thus there may be some discrepancies for those who do not round off the value of $\sin 25°$ from 0.4226182617 to 0.4226, for example. There may also be differences in answers when dealing with radian measure of angles. For example, $\sin 7.00$ is 0.6570 correct to four decimal places, yet if the value of π is taken as 3.14, then $\sin 7.00$ is approximately equal to $\sin[(0.72 + 2(3.14)]$, which is equal to $\sin 0.72$, which turns out to be 0.6594.

I believe that the use of calculators should be encouraged—many students have been turned off in their study of trigonometry by the tedious calculations required to obtain "practical" answers. Having access to a scientific calculator, however, does not make one an expert in trigonometry. Students need to know *why* the light flashes or the error sign shows when they attempt to find the angle whose sine is 2. They need to know that $(\sin 50°)^2 + (\cos 50°)^2$ is equal to exactly 1, rather than 0.99994784 or some other approximation obtained by squaring and adding values from the tables. Careful attention to the definitions and interrelations given in this book will provide a solid foundation for future work in mathematics and for applications in a wide variety of technical and scientific studies.

FRANK C. DENNEY
Hayward, California
November 1975

Acknowledgments

I wish to thank a number of persons who assisted in the preparation of this text. Dale E. Boye of Schoolcraft College, Robert Kester of East Los Angeles College, John L. Lawson of Cuyahoga Community College, Lawrence A. Trivieri of Mohawk Valley Community College, Jay Welch of San Jacinto Community College, and Ruth W. Wing of Palm Beach Junior College read the text and provided thoughtful comments and suggestions for improvement of its early versions. My friend and colleague from Chabot College, Pamela Matthews, class-tested a preliminary edition of this text and also helped shape the final form of the book. Ann Ludwig, mathematics editor of Canfield Press, gave great encouragement in getting the project started, while Pat Brewer, production editor, provided invaluable assistance and expertise in completing it. Finally, credit should go to Esperanza Bayol for her rapid but accurate typing of the manuscript, without which the book would not have been possible.

F.C.D.

TRIGONOMETRY

CHAPTER ONE

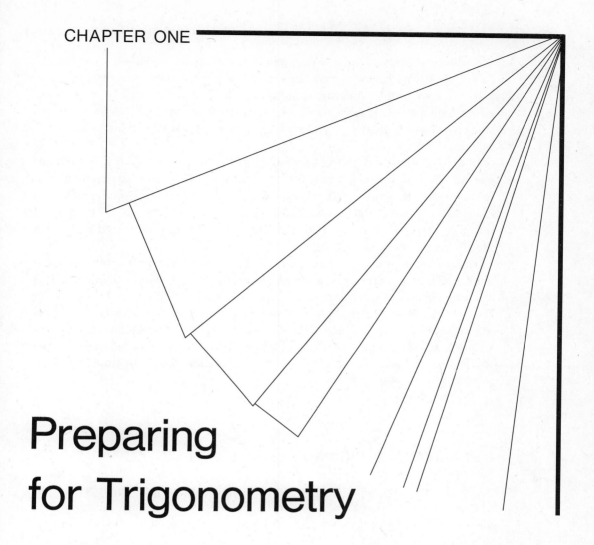

Preparing for Trigonometry

1.1 INTRODUCTION

Students who undertake to study trigonometry for the first time are at an exciting stage of their mathematical development. Having studied arithmetic, algebra, and geometry, they are ready to combine the knowledge gained in each of these subjects, add some new definitions and techniques, and move forward into a vastly increased range of applications and understandings.

The beginnings of what we call trigonometry are lost in antiquity. Trigonometric tables have been found on Babylonian tablets which date back to somewhere between 1900 and 1600 B.C. The Greek astronomer Hipparchus, who lived around 140 B.C., used trigonometry in his observations and was able to determine the distance from the earth to the moon, as well as the moon's circumference and diameter. The method used by Hipparchus was extremely simple, and you will be able to understand it by the end of the next section, but the calculation of the tables and ratios required was quite tedious and subject to considerable error.

We have progressed quite a long way since the days of Hipparchus and the early Greeks who helped lay the foundations of mathematics. Contributions have been made by people from throughout the world for centuries. We use and have used trigonometry for navigation, architecture, surveying, and as a basis for our study of advanced mathematics, physics, engineering, electronics and many other subjects.

In order for the student to assimilate in a few short months much of what has been developed over such a long period of time, it will be necessary to read carefully, listen well, and to work homework diligently, consistently and neatly. It is frequently observed that mathematics is not a spectator sport. The study of trigonometry is no different. It will be hard work, but enjoyable and well worth the effort to those who succeed.

There are no exercises at the end of this introductory section, but those who are at all interested are urged to consult a book on the history of mathematics, such as *An Introduction to the History of Mathematics*, revised edition, by Howard Eves (New York: Holt, Rinehart, and Winston, 1964), or on applications of mathematics, such as *Mathematics and the Physical World*, by Morris Kline (New York: Crowell, 1959). It might also be interesting to check out another trigonometry text in order to see how different and yet at the same time how similar two approaches to the same subject can be.

1.2 RIGHT TRIANGLE TRIGONOMETRY

Right triangle trigonometry is so simple that its importance might tend to be underestimated. All that is necessary is familiarity with some concepts from geometry, knowledge of some easy definitions, and the ability to use a set of tables of trigonometric ratios. Those of you who have forgotten the geometrical terms used in this section may wish to read through Section 1.4 before going ahead.

Consider a right triangle ABC with right angle at C. If A is one of the acute angles, then BC is the SIDE OPPOSITE angle A, AC is the SIDE ADJACENT to angle A, and AB, the side opposite the right angle, is the HYPOTENUSE, as shown in Figure 1–1. We define the following ratios:

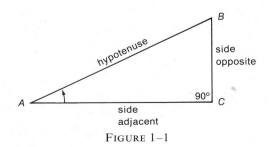

FIGURE 1–1

1.2 RIGHT TRIANGLE TRIGONOMETRY

The SINE of A is the ratio of the side opposite A to the hypotenuse.

The COSINE of A is the ratio of the side adjacent A to the hypotenuse.

The TANGENT of A is the ratio of the side opposite A to the side adjacent A.

The COTANGENT of A is the ratio of the side adjacent A to the side opposite A.*

The ratios are generally abbreviated by using the first three letters of the name so that we can write

$$\sin A = \frac{\text{side opposite}}{\text{hypotenuse}} \qquad \cos A = \frac{\text{side adjacent}}{\text{hypotenuse}}$$

$$\tan A = \frac{\text{side opposite}}{\text{side adjacent}} \qquad \cot A = \frac{\text{side adjacent}}{\text{side opposite}}$$

We also frequently use what is known as STANDARD NOTATION for lettering the sides and angles of a triangle, in which a lower case letter is used to denote the length of the side opposite the angle labeled with the corresponding upper case letter, as in Figure 1–2. Here we see that

$$\sin A = \frac{a}{c} \qquad \cos A = \frac{b}{c} \qquad \tan A = \frac{a}{b} \qquad \cot A = \frac{b}{a}$$

We can also note that

$$\sin B = \frac{b}{c} \qquad \cos B = \frac{a}{c} \qquad \tan B = \frac{b}{a} \qquad \cot B = \frac{a}{b}$$

It therefore follows that

$$\sin A = \cos B \qquad \cos A = \sin B \qquad \tan A = \cot B \qquad \cot A = \tan B$$

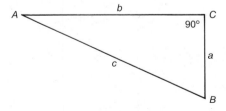

FIGURE 1–2

---

* There are two additional ratios, the secant and cosecant, which we shall introduce in Section 2.1.

These equivalents are important when using tables of trigonometric ratios. The sum of the measures of angles A and B is 90 degrees, hence A and B are complementary angles. The basis for the names *co*sine and *co*tangent is evident, as they equal the sine and tangent of the complementary angles. The pairs sine-cosine and tangent-cotangent are sometimes called COFUNCTIONS.

Since $A + B = 90°$, $B = 90° - A$ and $A = 90° - B$. Thus we find that $\sin 30° = \cos 60°$, $\tan 50° = \cot 40°$, and so on. Tables of values for the trigonometric ratios have been calculated and are provided in Table 1 of Appendix D. A few of the values are easy to determine and are exact, as will be shown in the next example, but most are quite difficult to evaluate and their approximate values are given, correct to four decimal places.

Consider the 30°, 60°, right triangle in Figure 1–3, with hypotenuse of length 2. From geometry we know that the side opposite the 30° angle is half the length of the hypotenuse, so that $b = 1$. Using the Pythagorean relation, $a^2 + b^2 = c^2$, we can calculate a value for a.

$$a^2 + 1^2 = 2^2$$

$$a^2 = 3$$

$$a = \sqrt{3}.$$

Thus by means of the definitions we find that

$$\sin 60° = \frac{\sqrt{3}}{2} \qquad \cos 60° = \frac{1}{2} \qquad \tan 60° = \sqrt{3} \qquad \cot 60° = \frac{1}{\sqrt{3}} = \frac{1}{3}\sqrt{3}$$

and

$$\sin 30° = \frac{1}{2} \qquad \cos 30° = \frac{\sqrt{3}}{2} \qquad \tan 30° = \frac{1}{\sqrt{3}} = \frac{1}{3}\sqrt{3} \qquad \cot 30° = \sqrt{3}$$

Referring to Table I in Appendix D, we see that $\sin 30° = 0.5000$, $\cos 30° = 0.8660$, $\tan 30° = 0.5774$, and $\cot 30° = 1.732$. The first of these values happens to be exact, but since $\sqrt{3}$ is equal to 1.732051, rounded to six decimal places,

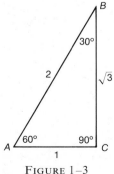

FIGURE 1–3

1.2 RIGHT TRIANGLE TRIGONOMETRY

$\sqrt{3}/2$ is more nearly 0.866025 than 0.8660. The table gives only an approximate value for cos 30°, and also for tan 30° and cot 30°.

When you look for the values of the trigonometric ratios for a 60° angle, you will find that the table seems to end after the ratios for 45°. This is because the tables are arranged to take advantage of the cofunction relationships. Any acute angle greater than 45° can be expressed in terms of an angle less than 45° by subtracting it from 90°, that is, 90° − 80° = 10°, 90° − 60° = 30°, and so on. We proved earlier that cos 60° = sin (90° − 60°) = sin 30°. Looking at the angles in the right hand column and using the column labels at the foot of the column, we find that cos 60° = sin 30° = 0.5000, sin 60° = 0.8660, cot 60° = 0.5774 and tan 60° = 1.732.*

Before leaving the 30°, 60°, right triangle for the moment, let us consider one with hypotenuse of length 1, as in Figure 1–4. Then the side opposite the 30° angle is 1/2, and the side opposite the 60° angle is $\sqrt{3}/2$, since

$$a^2 + \left(\frac{1}{2}\right)^2 = 1^2$$

$$a^2 = \frac{3}{4}$$

$$a = \frac{\sqrt{3}}{2}$$

Again by means of the definitions, we find

$$\sin 60° = \frac{(\sqrt{3}/2)}{1} = \sqrt{3}/2$$

$$\cos 60° = \frac{1/2}{1} = \frac{1}{2}$$

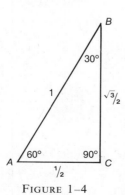

FIGURE 1–4

* Recall that tan 60° is *exactly* equal to $\sqrt{3}$, and only *approximately* equal to 1.732. We shall use the equal sign for convenience, however, while noting that there is a difference.

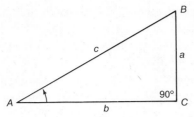 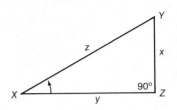

FIGURE 1–5

$$\tan 60° = \frac{(\sqrt{3}/2)}{(1/2)} = \sqrt{3}$$

$$\cot 60° = \frac{(1/2)}{(\sqrt{3}/2)} = \frac{1}{\sqrt{3}} = \frac{1}{3}\sqrt{3}$$

exactly the same as before, since all right triangles with a given acute angle are similar, and their corresponding sides are proportional. If a right triangle ABC with right angle at C has the measure of angle A equal to the measure of angle X, an angle of right triangle XYZ with right angle at Z, as in Figure 1–5, we know from geometry that the triangles are similar and their corresponding sides are proportional. In other words,

$$\frac{a}{x} = \frac{b}{y} = \frac{c}{z}$$

If $x = ka$, $y = kb$, and $z = kc$, then

$$\sin X = \frac{x}{z} = \frac{ka}{kc} = \frac{a}{c} = \sin A$$

$$\cos X = \frac{y}{z} = \frac{kb}{kc} = \frac{b}{c} = \cos A$$

$$\tan X = \frac{x}{y} = \frac{ka}{kb} = \frac{a}{b} = \tan A$$

The values of the trigonometric ratios are dependent on the size of the angle, not the size of the triangle of which it is part.

Now let us apply the trigonometric ratios to the solution of some problems. The list of different applications is nearly endless, but it will be evident that your imagination will enable you to use the same basic ideas in the solution of all of them.

1.2 RIGHT TRIANGLE TRIGONOMETRY

EXAMPLE 1   Suppose that you want to find the height of an object above a point on the ground. In this case you are in a weather station at point W on the ground and there is a searchlight at S, 1000 feet away, pointing straight up toward the clouds where it reflects at point C. You measure angle SWC and find it to be 50°. Let w be the height of the clouds, as indicated in Figure 1-6.

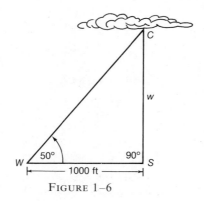

FIGURE 1-6

SOLUTION   The side opposite angle W and the side adjacent are known, so use of the tangent ratio is indicated. By the definition, $\frac{w}{1000} = \tan 50°$, or $w = 1000 \tan 50°$. From Table I in Appendix D we find that $\tan 50° = 1.192$ (remember to use the column label at the bottom of the table for angles greater than 45°), so $w = 1000 \,(1.192) = 1192$, and the height of the clouds is about 1190 feet.*

EXAMPLE 2   A brace 15.0 cm long is used to strengthen part of an electronics assembly. One end of the brace is soldered 8.0 cm from one corner. How far from the corner is the other end soldered, and what angles does it make with the sides?

SOLUTION   If we draw a sketch as in Figure 1-7, we see that we need to find side a, angle A, and angle B. We could find side a by means of the Pythagorean relation, but let us use the trigonometric ratios instead. Since AB is the hypotenuse and AC is the side adjacent to angle A, we can write $\cos A = \frac{8.0}{15.0}$, or $\cos A = 0.5333$, correct to 4 decimal places. Looking in the body of Table I in Appendix D, we find that $\cos 57°50' = 0.5324$,

---

* A word about accuracy: We shall assume that our measures are exact, and we shall carry out calculations to three significant digits for lengths of sides of triangles, or to the nearest ten minutes for the measures of angles. Most of our calculations will have been performed using an electronic calculator with a capacity of 8 digits or more, and the tables in this text.

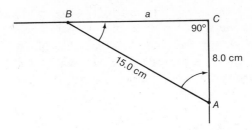

FIGURE 1-7

which is reasonably close to 0.5333 (nearer than 0.5348, the value of cos 57°40′). Since A and B are complements, $B = 90° - 57°50' = 32°10'$.*
Then $\dfrac{a}{15.0} = \sin 57°50'$, or $a = 15.0 \sin 57°50' = 15.0\,(0.8465) = 12.7$, correct to 3 significant digits.

EXAMPLE 3   A VECTOR is a line segment that is used to represent a force, velocity, or acceleration moving in a certain direction. One operation on vectors is known as "resolving the vector into its horizontal and vertical components." This involves finding the lengths of the sides labeled x and y in a right triangle as indicated in Figure 1-8. Here we see a vector which has a "magnitude" of 30 and a "direction angle" of 40°. Find x and y.

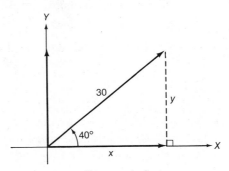

FIGURE 1-8

SOLUTION   Since y is the side opposite the 40° angle and x is the side adjacent, $\dfrac{x}{30} = \cos 40°$ and $\dfrac{y}{30} = \sin 40°$, so $x = 30 \cos 40° = 30\,(0.7660) = 23.0$ and $y = 30 \sin 40° = 30\,(0.6428) = 19.3$, both to 3 significant digits.

In some applications it is useful to have the terms angle of elevation and angle of depression in order to refer to the acute angles formed by a horizontal line and the line of sight from an observer to an object. In Figure 1-9 if a passenger in a helicopter at point H sees a person on the ground at point G, then angle KHG is the ANGLE OF DEPRESSION, while angle FGH is the ANGLE OF ELEVATION. Note that these angles would have the same measure.

* Note that 32°10′ is the angle at the left hand side of the table when finding angle A.

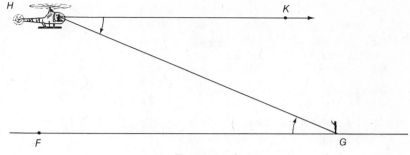

FIGURE 1-9

EXERCISES

1. Find the value of each of the following in Table I in Appendix D:
a. sin 32°
b. cos 58°
c. tan 29°20′
d. cot 60°40′
e. cos 49°50′
f. cot 35°10′
g. sin 10°10′
h. tan 83°20′

2. Find the value of each of the following in Table I in Appendix D:
a. cos 41°
b. sin 49°
c. cot 72°30′
d. tan 17°30′
e. sin 88°50′
f. tan 88°50′
g. sin 1°10′
h. tan 1°10′

3. Find the angle A to the nearest ten minutes, using Table I in Appendix D:
a. tan $A = 0.6009$
b. cot $A = 0.6009$
c. sin $A = 0.7214$
d. cos $A = 0.7214$
e. sin $A = 0.0080$
f. cos $A = 0.6840$
g. cot $A = 1.900$
h. tan $A = 2.687$

4. Find the angle B to the nearest ten minutes, using Table I in Appendix D:
a. cos $B = 0.4041$
b. sin $B = 0.4041$
c. cot $B = 17.17$
d. tan $B = 17.17$
e. cos $B = 0.9691$
f. cot $B = 0.5290$
g. sin $B = 0.7906$
h. tan $B = 0.5600$

5. Draw a right triangle with legs of length 2.
a. Find the exact length of the hypotenuse.
b. Calculate the exact values of the sine, cosine, tangent, and cotangent of a 45° angle.
c. Compare the values obtained in part b with the values given in Table I in Appendix D. State which are exact.

6. Draw a right triangle with hypotenuse of length 4 and one leg of length 2.
a. What is the exact length of the third side of the triangle?
b. What are the measures of the acute angles of the triangle?
c. Determine the exact values of the sine, cosine, tangent, and cotangent of each of the acute angles by means of the definitions of the trigonometric ratios.
d. Compare the results obtained in part c with the results on page 4.

In each of the following problems, sketch the triangle and label the given parts. Use trigonometric ratios where possible, although you may wish to check by means of the Pythagorean relation. Round answers to 3 significant digits or to the nearest 10 minutes.

7. Given right triangle ABC with $A = 29°$, $C = 90°$, and side c of length 20, find
a. angle B
b. side a
c. side b.

8. Given right triangle ABC with $A = 53°40′$, $C = 90°$, and side c of length 35, find
a. angle B
b. side a
c. side b.

9. Given right triangle XYZ with $Z = 90°$, side $x = 25$ and side $y = 18$, find
a. angle X
b. angle Y
c. hypotenuse z.

10. Given right triangle PQR with $R = 90°$, side $q = 16.8$ and hypotenuse $r = 30$, find
a. angle P
b. angle Q
c. side p.

11. The sides of an isosceles triangle are 8.0 meters, 8.0 meters, and 9.6 meters, respectively. Find a. the length of the altitude to the 9.6 meter base b. the measures of the angles c. the area of the triangle.

12. The two equal sides of an isosceles trapezoid are each 12 inches long, and they form angles of 65° with the base, which is 20 inches long. Find a. the altitude of the trapezoid b. the length of the upper base c. the area of the trapezoid.

EXAMPLE 4 Find the perimeter and area of a regular octagon inscribed in a circle with radius 3.5 feet.

SOLUTION The octagon with its diagonals is shown in Figure 1–10. The central angles of a regular polygon, such as angle BOC, are all equal, so $m°(\angle BOC) = m°(\angle AOB) = 360°/8 = 45°$. If a perpendicular OD is drawn from the center O to side AB, OD BISECTS AB or divides it into two equal parts so that $AD = DB$, and also bisects angle AOB, so that $m°(\angle BOD) = m°(\angle AOD) = 45°/2 = 22°30'$.

To find the perimeter of the octagon we must find the length of AB and multiply it by 8: $P = 8AB$. If $x = AD$, then $AB = 2x$, so $P = 8(2x) = 16x$. In the right triangle AOD, $\sin 22°30' = x/3.5$, so $x = 3.5 \sin 22°30' \approx 3.5(0.3827) \approx 1.339$.* Thus $P \approx 16(1.339) \approx 21.4$ feet.

To find the area of the octagon we need to find the area of triangle AOB and multiply it by 8: $\mathscr{A} = 8(1/2)(AB)(OD)$. If we let $y = OD$, and $x = AD = (1/2)(AB)$, then $\mathscr{A} = 8xy$. In the right triangle AOD, $\cos 22°30' = y/3.5$, so $y = 3.5 \cos 22°30' \approx 3.5(0.9239) \approx 3.234$. Thus $\mathscr{A} \approx 8(1.339)(3.234) \approx 34.6$ square feet.

13. Find the perimeter and the area of a regular pentagon inscribed in a circle with radius 3.03 mm.

14. Show that the area of a regular polygon of n sides inscribed in a circle with radius r can be found by means of the formula

$$\mathscr{A} = nr^2 \sin\left(\frac{180°}{n}\right) \cos\left(\frac{180°}{n}\right)$$

15. A helicopter flies due north from an airport for 30 miles, then turns and heads due east. A radar operator at the airport observes a few minutes later that the helicopter is located 20° east of the north-south line. What is the distance of the helicopter from the airport at that time?

16. A balloon is released from a weather station and rises at the steady rate of 100 feet per minute. Ten minutes later observers at the station note that the angle of elevation of the balloon is 22°30'. Over what horizontal distance has the balloon traveled, and what does this indicate the wind velocity to be in miles per hour?

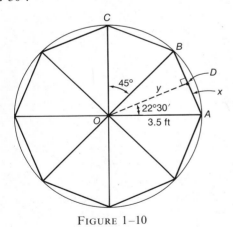

FIGURE 1–10

1.3 A PREVIEW OF THINGS TO COME

The trigonometric ratios defined in the previous section have an endless variety of applications. Since any polygonal figure can be broken up into a set of triangles,

* The sign $\approx$ is used when values are approximate.

1.3 A PREVIEW OF THINGS TO COME

it is possible to compute the area or diagonal distances by this decomposition process. Special formulas for comparatively rapid computation of distances can also be developed.

EXAMPLE 1   Given triangle RST with $RS = 500$, angle $R = 40°$ and angle $S = 25°$, find the altitude TU, from T to RS (Refer to Figure 1–11).

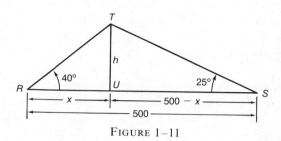

FIGURE 1–11

SOLUTION   We can resort to a little bit of algebra here, and let $x = RU$, $500 - x = SU$, and $h = TU$. Then

$$\cot 40° = \frac{x}{h}$$

$$x = h \cot 40°$$

$$\cot 25° = \frac{500 - x}{h}$$

$$500 - x = h \cot 25°$$

Now we have two equations in two variables, x and h. If we substitute $x = h \cot 40°$ into the second equation, we obtain $500 - h \cot 40° = h \cot 25°$, so $500 = h \cot 25° + h \cot 40°$. Thus

$$500 = h(\cot 25° + \cot 40°)$$

$$h = \frac{500}{\cot 25° + \cot 40°}$$

If we need a numerical approximation for some practical purpose we can look up the ratios in Table I in Appendix D to find

$$h = \frac{500}{2.145 + 1.192} \approx 150$$

By substituting the unknown d for RS, a formula which would handle all similar circumstances can be written

$$h = \frac{d}{\cot R + \cot S}$$

You may wonder why it is necessary to have a quarter or semester course with a text of over 200 pages if the relatively simple material just covered is so applicable and powerful. Although it *is* true that with a few more days and the equivalent of one chapter of this text, specifically the last chapter, in which the Law of Sines and the Law of Cosines are introduced, it would be possible to handle a significant portion of technical, engineering, and physical applications, there is still much more to modern trigonometry.

We live in a world not only of static things in which lengths and distances are to be measured, but also of moving things which go around in circles, which appear and disappear in cycles, and for which analysis by trigonometric means is possible if we generalize the definitions of the trigonometric ratios slightly. In Chapter Two we will do this, and learn how to find $\sin 2000°$ or $\cos(-135°)$, making a distinction between counterclockwise rotations, which are considered positive, and clockwise rotations, which are then negative.

In Chapter Three we will work out a great many interrelationships among our newly defined trigonometric ratios. We have already seen that in a right triangle, for example, $\cos A = \sin(90° - A)$. The reader who has a calculator or the patience to do a little work in arithmetic with pencil and paper can verify that for any angle A in Table I in Appendix D it is true that $\tan A = \frac{\sin A}{\cos A}$, that $(\sin A)^2 + (\cos A)^2 = 1$, and that $\sin 2A = 2 \sin A \cos A$. In Chapter Three it will be proved that these relationships are true for all angles A, not just those in Appendix D. These formulas are useful in all kinds of settings, not only in advanced mathematics but also in the sciences and engineering.

Chapter Four is devoted to graphs of the trigonometric functions, which it turns out our trigonometric ratios can be. For example, if we plot the graph of $y = \sin x$, letting x take on values from $0°$ through $90°$, we obtain the darker portion of the graph in Figure 1–12. If we use the new definitions of Chapter Two, the graph can be extended infinitely far to the left or right as the figure indicates.

Many of you have encountered the complex numbers in algebra. In Chapter Five we review the algebraic notions, then define them in terms of trigonometric functions and learn how to multiply, divide, and take powers and roots of complex

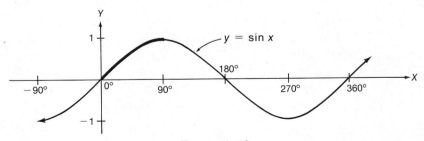

Figure 1–12

1.3 A PREVIEW OF THINGS TO COME

numbers in a relatively simple way. We shall also use a polar coordinate system for locating points, which will complement the rectangular coordinate system we have used previously.

Finally in Chapter Six we get everything together in order to investigate a wide range of practical applications, from the solution of triangles to the cyclic phenomena mentioned earlier. As we proceed through the course, an application will be demonstrated or hinted at wherever possible, but sometimes a truly practical application requires a combination of facts and skills, which is the reason for our last chapter. For the time being, some additional practical applications of right triangles are found in the exercises that follow.

EXERCISES

Use the values in Table I in Appendix D to show that each of the following statements is true. (There may be slight differences due to rounding off. Work to three significant digits.)

1. $\tan 40° = \dfrac{\sin 40°}{\cos 40°}$

2. $\cot 20° = \dfrac{\cos 20°}{\sin 20°}$

3. $(\sin 55°)^2 + (\cos 55°)^2 = 1$

4. $(\cos 35°)^2 - (\sin 35°)^2 = \cos 70°$

5. $\sin 30° = \sqrt{\dfrac{1 - \cos 60°}{2}}$

6. $\tan 20° = \dfrac{\sin 40°}{1 + \cos 40°}$

7. $\sin 50° \neq \sin 30° + \sin 20°$ (*Note:* The sign $\neq$ means "is not equal.")

8. $\sin 50° = \sin 30° \cos 20° + \cos 30° \sin 20°$

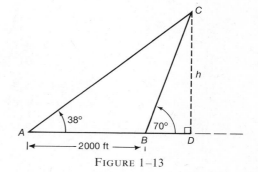

FIGURE 1–13

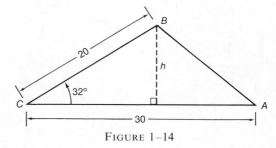

FIGURE 1–14

9. Two observers on the ground 2000 feet apart see an object in the air between them. If the angle of elevation from one observer is 70° and from the other is 38°, find the height of the object.

10. Two observers on the ground 2000 feet apart see an object in the air and not between them, as in problem 9, but beyond the one for which the angle of elevation is 70°, as shown in Figure 1–13. Find the height of the object.

11. Find the area of the triangle shown in Figure 1–14. *Hint:* First determine h.

12. Show that if a and b are the sides opposite angles A and B, respectively, in Figure 1–14, the area $\mathscr{A}$ of triangle ABC can be found by means of the formula $\mathscr{A} = (1/2)ab \sin C$.

13. If the straight line distance between two observatories located at points A and B on the earth is 1390 miles (Figure 1–15), $CE = 62$ miles, and angle MAB = angle ABM = 89°50′, find the distance EM from the earth to the moon.

14. In a surveying class Ed found the measure of angle RPQ in a certain problem

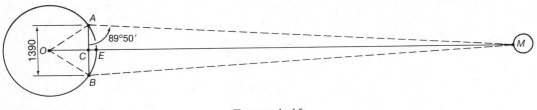

FIGURE 1–15

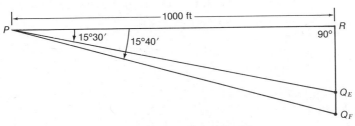

FIGURE 1–16

(Figure 1–16) to be 15°30′ while Fred found it to be 15°40′. If angle PRQ is 90° and $PR = 1000$ ft, what is the difference between RQ_E (Ed's calculation of the distance RQ) and RQ_F (Fred's calculation of the distance RQ)?

1.4 FACTS AND FORMULAS FROM GEOMETRY

After studying the previous two sections you should be convinced that there is much which must be remembered from geometry in order to learn trigonometry. The following is a summary of some of the basic ideas needed in the remainder of the text.

ANGLES are composed of two rays (parts of a line) which have a common endpoint called the VERTEX of the angle, as in Figure 1–17. The rays are called the SIDES of the angle. In Figure 1–17 the vertex is point O, and rays OA and OB are the sides of the angle. We can also call a LINE SEGMENT (the set of points between and including two distinct points) the side of an angle. Segment OA of Figure 1–17 consists of points O and A together with all points of line OA between O and A, while ray OA consists of segment OA and all points of line OA beyond

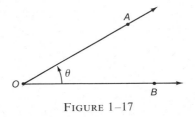

FIGURE 1–17

1.4 FACTS AND FORMULAS FROM GEOMETRY

point A. We can name an angle by means of a single letter, as in angle O of Figure 1–17, or we may choose to use three letters, the vertex and a point on each side, as in angle AOB. The vertex letter is always placed between the other two. It is often convenient to use a Greek letter to name an angle, and here we show the letter θ (theta) inside angle AOB. In Figure 1–18 it can be seen that a reference to angle E would be ambiguous, so in this case we would use angle CEF, FED, or CED as appropriate. Also angle CEF = angle α (Greek letter alpha) and angle FED = angle β (Greek letter beta).

Angles are measured by various means, but you are probably most familiar with degree measure. A DEGREE is the unit of measure of an angle whose rays intercept an arc which is 1/360 part of the circumference of a circle. For more precise measurements the degree can be divided into sixty equal parts called MINUTES, so that 1 degree = 60 minutes ($1° = 60'$), and the minute can be divided into 60 equal parts called SECONDS, so that 1 minute = 60 seconds ($1' = 60''$). The possible confusion of symbols for minutes and seconds with the symbols for feet and inches can be avoided by always using the degree symbol in conjunction with those for minutes and seconds, as in $0°25'$ to refer to a 25 minute angle.

A STRAIGHT ANGLE is an angle whose sides are opposite rays, as in Figure 1–19. Angle PQR has sides QP and QR, and vertex at Q. Angle PQR measures $180°$. A RIGHT ANGLE measures $90°$, an ACUTE ANGLE measures between $0°$ and $90°$, and an OBTUSE ANGLE measures between $90°$ and $180°$. When two lines intersect, as in Figure 1–20, the opposite pairs of angles, called VERTICAL ANGLES, are equal in measure: $m°(\angle AOC) = m°(\angle DOB)$ and $m°(\angle COB) = m°(\angle DOA)$. If the sum of the measures of two angles is $180°$, the angles are said to be SUPPLEMENTARY ANGLES. Since $m°(\angle AOC) + m°(\angle DOA) = m°(\angle DOC) = 180°$ in Figure 1–20, angles AOC and DOA are supplementary angles. Supplementary angles do not have to have the same vertex. If $m°(\angle X) + m°(\angle Y) = 180°$, then angle X and angle Y are supplementary. If the sum of the measures of two angles is $90°$, the angles are said to be COMPLEMENTARY ANGLES. Referring to Figure 1–21, we see that angle R and angle T are supplementary, since $m°(\angle R) +$

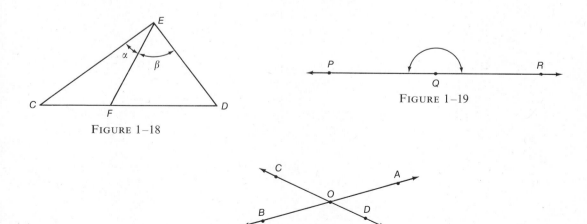

FIGURE 1–18

FIGURE 1–19

FIGURE 1–20

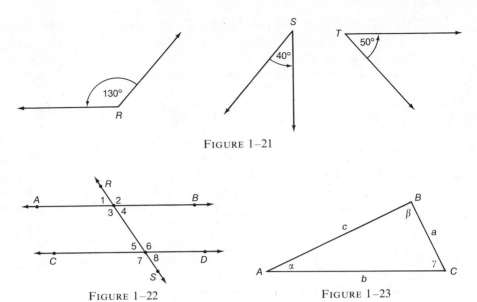

FIGURE 1–21

FIGURE 1–22   FIGURE 1–23

$m°(\angle T) = 180°$, while angle S and angle T are complementary, since $m°(\angle S) + m°(\angle T) = 90°$.

If two parallel lines are intersected by a third line, called a TRANSVERSAL, certain pairs of angles formed are equal in measure or are supplementary. In Figure 1–22 parallel lines AB and CD are intersected by transversal RS. The angles 3 and 6, 4 and 5 are called ALTERNATE-INTERIOR ANGLES, and their measures are equal: $m°(\angle 3) = m°(\angle 6)$ and $m°(\angle 4) = m°(\angle 5)$. Angles 2 and 6, 4 and 8, 1 and 5, 3 and 7 are called CORRESPONDING ANGLES, and their measures are equal: $m°(\angle 2) = m°(\angle 6)$, $m°(\angle 4) = m°(\angle 8)$, $m°(\angle 1) = m°(\angle 5)$, and $m°(\angle 3) = m°(\angle 7)$. Angles 4 and 6, 3 and 5 are called INTERIOR ANGLES and they are supplementary: $m°(\angle 4) + m°(\angle 6) = 180°$ and $m°(\angle 3) + m°(\angle 5) = 180°$.

TRIANGLES are sets of points formed by line segments which join three distinct points not on the same line. The line segments are the sides of the three angles of the triangle and the three distinct points are the vertices (plural of vertex) of the triangle. In triangle ABC of Figure 1–23, points A, B, and C are the vertices and segments AB, BC, and CA are the sides. It is frequently convenient to denote the length of the side opposite an angle by a lower case letter corresponding to the letter of the vertex point, and to use Greek letters as mentioned before to denote the angles. In Figure 1–23 the Greek letter gamma (γ) denotes angle C. This standard notation will be used frequently in examples and exercises involving triangles.

Triangles can be classified by their angles as acute, right, or obtuse triangles. An ACUTE TRIANGLE has all three angles acute, a RIGHT TRIANGLE has one right angle, and an OBTUSE TRIANGLE has one obtuse angle. The sum of the measures of the three angles of a triangle is exactly 180°, so that in Figure 1–23, $m°(\angle \alpha) + m°(\angle \beta) + m°(\angle \gamma) = 180°$.

1.4 FACTS AND FORMULAS FROM GEOMETRY

Triangles can be classified by the lengths of their sides as scalene, isosceles, or equilateral triangles. A SCALENE TRIANGLE has all three sides of different length, an ISOSCELES TRIANGLE has two sides of equal length, and an EQUILATERAL TRIANGLE has three sides of equal length. The sum of the lengths of any two sides is always greater than the length of the third side.

Certain relationships between sides and angles can be useful. In any triangle the side of greatest length is always opposite the angle of greatest measure, the side of least length is always opposite the angle of least measure. In Figure 1–24, for example, $x < z < y$ since $m°(\angle X) < m°(\angle Z) < m°(\angle Y)$ and $m°(\angle S) < m°(\angle R) < m°(\angle T)$ since $10 < 12 < 15$. In an isosceles triangle the angles opposite the sides of equal length are equal in measure, and in an equilateral triangle the angles are all of equal measure (which means that each angle of an equilateral triangle measures 60°).

In discussing right triangles, special terms are used to describe the relationships of angles and sides. In right triangle ABC of Figure 1–25, in which $m°(\angle C) = 90°$, AB, the side opposite the right angle, is called the hypotenuse, while AC and BC, the other two sides, are called the legs. The PYTHAGOREAN RELATION will always hold for a right triangle: the square of the length of the hypotenuse is equal to the sum of the squares of the lengths of the legs ($c^2 = a^2 + b^2$). The acute angles of a right triangle are complementary.

An isosceles right triangle (Figure 1–26) has legs of equal length, so that $a = b$, and $m°(\angle A) = m°(\angle B) = 45°$. From the Pythagorean relation we derive $c^2 = a^2 + b^2 = a^2 + a^2 = 2a^2$, and if $c^2 = 2a^2$, $c = a\sqrt{2}$.

A 30°, 60°, right triangle is important in trigonometry because of a special relationship among the sides. This relationship can be seen in the construction

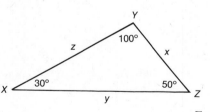
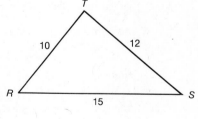

FIGURE 1–24

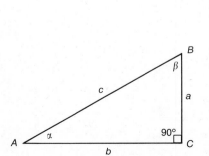

FIGURE 1–25

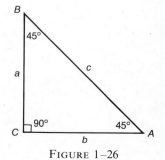

FIGURE 1–26

of Figure 1-27. The ALTITUDE of a triangle is a segment drawn from a vertex perpendicular (forming a right angle) to the base. If an altitude is drawn from a vertex of an equilateral triangle, two congruent right triangles are formed (CONGRUENT TRIANGLES have corresponding angles and corresponding sides of equal measure). Thus we see in Figure 1-27 that altitude BC of triangle ABD divides it into congruent triangles ABC and DBC, so that $m°(\angle ABC) = m°(\angle DBC) = 30°$, and

$$x = AC = CD = \frac{1}{2}(AD) = \frac{1}{2}(AB) = \frac{1}{2}d$$

Thus the side opposite the 30° angle in a 30°, 60°, right triangle is half the length of the hypotenuse. Also $h^2 + x^2 = d^2$, or

$$h^2 = d^2 - x^2 = d^2 - \left(\frac{1}{2}d\right)^2 = d^2 - \frac{1}{4}d^2 = \frac{3}{4}d^2$$

and if $h^2 = \frac{3}{4}d^2$, $h = \frac{1}{2}\sqrt{3}\,d$. The side opposite the 60° angle in a 30°, 60°, right triangle is $\frac{1}{2}\sqrt{3}$ times the length of the hypotenuse.

The AREA OF A TRIANGLE is equal to one-half the product of the lengths of the base and the altitude to that base, which can be denoted by $\mathscr{A} = (1/2)bh$. (The script letter $\mathscr{A}$ is used to distinguish area from a point or vertex.)

SIMILAR TRIANGLES are triangles which have two angles of one equal in measure to two angles of the other. (If two angles are equal, the third angle *must* be equal, since the sum of the angles of a triangle is 180°.) The corresponding sides of similar triangles are proportional. In Figure 1-28, triangles ABC and DEF are similar (what is the measure of angles C and F?) and $\frac{a}{d} = \frac{b}{e} = \frac{c}{f}$.

A POLYGON is a set of points formed by joining n distinct points by n line segments, none of which intersect except at their endpoints. A REGULAR POLYGON is a polygon with all of its sides of equal length and all of its angles of equal measure. The five-sided regular PENTAGON of Figure 1-29 is comprised of the

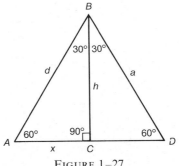

FIGURE 1-27

1.4 FACTS AND FORMULAS FROM GEOMETRY

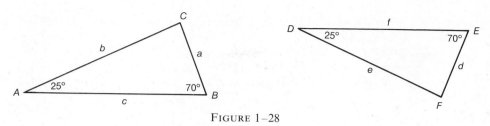

FIGURE 1-28

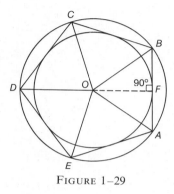

FIGURE 1-29

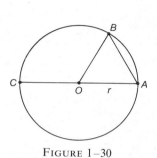

FIGURE 1-30

line segments or sides AB, BC, CD, DE, and EF. Its CENTRAL ANGLES AOB, BOC, COD, DOE, and EOA all have the same measure, so $m°(\angle AOB) = \frac{1}{5}(360°) = 72°$. If the regular polygon has n sides, a central angle measures $(360°/n)$. The circle with radius OA is CIRCUMSCRIBED about the polygon and the polygon is INSCRIBED in the circle. The circle with radius OF is inscribed in the polygon and the polygon is circumscribed about the circle. All radii of a circle are of equal length, so triangles AOB, BOC, and so on are isosceles triangles, and if an altitude is drawn from the center to a side, it divides the side into segments of equal length ($BF = FA$), and $m°(\angle FOB) = m°(\angle AOF)$. The PERIMETER of a polygon is the sum of the lengths of its sides.

A CIRCLE is a set of points equidistant from a fixed point called the CENTER of the circle. A segment joining the center to any point on the circle is called a RADIUS of the circle. The length of a radius is also referred to as the radius of the circle. A segment with endpoints on the circle which contains the center is called a DIAMETER of the circle. Referring to Figure 1-30, the various parts of the circle are radii OA, OB, and OC, while AC is a diameter. The set of points of the circle between and including points A and B form an ARC of the circle. Line segment AB is a CHORD. The set of points bounded by arc AB, radius OB, and radius OA form a SECTOR of the circle, while the set of points bounded by line segment AB and arc AB form a SEGMENT of the circle.

The CIRCUMFERENCE of a circle is the distance around it, the length of the great arc from A to B to C and on around to A again in Figure 1-30. The number π (Greek letter pi) is the ratio of the circumference of a circle to its radius. If c

is the length of the circumference and d is the length of a diameter, then $\pi = \dfrac{c}{d}$, and $c = \pi d$. The value of π is approximately 3.1415926, but rounding off to 3.14 will be adequate for the computations in this text. The AREA OF A CIRCLE can be found by multiplying π times the square of the radius ($\mathscr{A} = \pi r^2$).

EXERCISES

1. Determine the number of degrees in each of the following angles of Figure 1–31 if $BD = CD = 10$, $m°(\angle DBC) = 25°$, and $m°(\angle A) = 100°$:
 a. $m°(\angle C)$ b. $m°(\angle BDC)$
 c. $m°(\angle ABD)$ d. $m°(\angle ADB)$

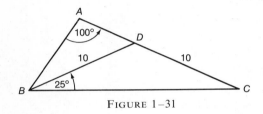

FIGURE 1–31

2. Determine the number of degrees in each of the following angles of Figure 1–32 if $RU = TU = 12$, $m°(\angle S) = 90°$, and $m°(\angle SUT) = 35°$:
 a. $m°(\angle UTS)$ b. $m°(\angle RUT)$
 c. $m°(\angle TRU)$ d. $m°(\angle RTS)$

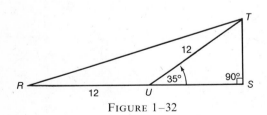

FIGURE 1–32

3. If parallel lines AB and CD of Figure 1–33 are cut by a transversal RS so that $m°(\angle ROB) = 2\,m°(\angle AOR)$, find the number of degrees in each of the following angles:
 a. $m°(\angle ROB)$ b. $m°(\angle RPD)$
 c. $m°(\angle RPC)$ d. $m°(\angle AOP)$

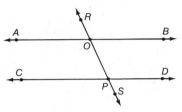

FIGURE 1–33

4. If parallel lines AB and CD of Figure 1–33 are cut by a transversal RS so that $m°(\angle OPD) = m°(\angle BOP) + 100°$, find the number of degrees in each of the following angles:
 a. $m°(\angle OPD)$ b. $m°(\angle BOP)$
 c. $m°(\angle CPO)$ d. $m°(\angle ROB)$

5. Find the length of the hypotenuse of a right triangle if the legs are of length
 a. 8 and 15 b. 7 and 24

6. Find the length of the hypotenuse of a right triangle if the legs are of length
 a. 10 and 24 b. 15 and $5\sqrt{7}$

7. Find the length of a leg of a right triangle if the hypotenuse is of length 17 and one leg measures 15.

8. Find the length of a leg of a right triangle if the hypotenuse is of length 15 and one leg measures $2\sqrt{14}$.

9. Find the lengths of the sides of a right triangle if the longer leg is 2 more than twice the length of the shorter leg and the hypotenuse measures 1 more than the longer leg.

10. Find the lengths of the sides of a right triangle if their lengths are three consecutive even integers.

1.5 FACTS AND FORMULAS FROM ALGEBRA

11. Find the exact perimeter and the area of a regular hexagon (6 sides) inscribed in a circle with radius 15 inches.

12. Find the exact perimeter and area of a regular hexagon (6 sides) circumscribed about a circle with radius 15 inches.

13. In Figure 1–34 triangles ABC and DEC are similar and DE is parallel to AB. If $DC = 10$, $DE = 6$, and $AB = 9$, find the lengths of AC and AD.

14. If the dimensions of the triangles in Figure 1–34 are the same as stated in problem 13 and if $BE = 4$, find the lengths of CE and CB.

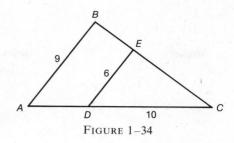

FIGURE 1–34

15. Find the exact perimeter and area of a square inscribed in a circle with radius 15 inches.

16. Find the exact area and perimeter of a square if its diagonal is 25 centimeters.

1.5 FACTS AND FORMULAS FROM ALGEBRA

It would take more space than is available here to review all of the algebra required for your study of trigonometry, but there are a few basic areas which tend to be forgotten which are listed for reference and review.

SPECIAL PRODUCTS AND FACTORING

PRODUCT OF TWO BINOMIALS: $(ax + b)(cx + d) = acx^2 + (ad + bc)x + bd$

SQUARE OF A BINOMIAL: $(a + b)^2 = a^2 + 2ab + b^2$

PRODUCT OF SUM AND DIFFERENCE: $(a + b)(a - b) = a^2 - b^2$

Each of the above can be written either way, that is $a^2 - b^2 = (a + b)(a - b)$. The problem is to recognize the type of expression involved and then to simplify the result.

EXAMPLE 1 Find the following products:
a. $(x + 3y)^2$ b. $(3x - 5)(3x + 5)$ c. $(2t + 3)(t - 7)$

SOLUTIONS a. This is the square of a binomial so $(x + 3y)^2 = x^2 + 2(x)(3y) + (3y)^2 = x^2 + 6xy + 9y^2$. b. This is the product of a sum and difference, since $(3x - 5)(3x + 5) = (3x + 5)(3x - 5)$, and $(3x + 5)(3x - 5) = (3x)^2 - 5^2 = 9x^2 - 25$. c. This is the product of two binomials so $(2t + 3)(t - 7) = 2t^2 + (-14 + 3)t - 21 = 2t^2 - 11t - 21$.

EXAMPLE 2   Factor each of the following: a. $(\cos T)^2 - (\sin T)^2$ b. $(\sin \alpha)^2 - 2(\sin \alpha)(\cos \alpha) + (\cos \alpha)^2$ c. $2u^2 - u - 3$

SOLUTIONS   a. This is the difference of two squares, which is the result of multiplying the sum times the difference of the same two numbers, so $(\cos T)^2 - (\sin T)^2 = (\cos T + \sin T)(\cos T - \sin T)$.   b. This is a perfect-square trinomial since the first and last terms are squares of numbers and the second term is twice the product of the numbers, so $(\sin \alpha)^2 - 2(\sin \alpha)(\cos \alpha) + (\cos \alpha)^2 = (\sin \alpha - \cos \alpha)^2$.   c. This is neither of the special cases of the previous two parts, so it is probably simply the product of two binomials. Since $2u^2 = 2(u)(u)$ and $3 = (3)(1)$, we could have products of the form $(2u\ \ 1)(u\ \ 3)$ or $(2u\ \ 3)(u\ \ 1)$. Since the coefficients of u must have an algebraic sum of 1, however, we choose the latter and determine the signs so that $(2u - 3)(u + 1) = 2u^2 - u - 3$. This type of factoring requires some trial and error work, but the exercises in this text will be fairly simple.

SETS OF NUMBERS

For the most part you will be working with the set of REAL NUMBERS in this book. The real number system is made up of sets of numbers described as follows:

The set of NATURAL NUMBERS, or counting numbers, is simply 1, 2, 3, 4, 5, and so on without end. We can denote this set by the letter N, and use the symbol $\in$ to mean "is a member of," so that $5 \in N$ means "5 is a member of the set N of natural numbers." The symbol $\notin$ means "is not a member of," so that $\pi \notin N$ means "π is not a member of the set N of natural numbers."

The set of INTEGERS includes the natural numbers, their negatives, and zero. We shall denote this set by the letter I, and point out that if $n \in N$, then $n \in I$ and $-n \in I$ but that $-n \notin N$. In other words the number 7 is a natural number, and both 7 and -7 are integers, but -7 is not a natural number. There will be many occasions when we will need to refer to integral multiples of a certain number. In this case we will let the letter k stand for the integer, and the fact that $k \in I$ will be understood. For example, $\alpha = 50° + k \cdot 90°$ means that α can be found by replacing k with any integer:

$$\alpha = 50° + 0 \cdot 90° = 50° + 0° = 50°$$
$$\alpha = 50° + 1 \cdot 90° = 50° + 90° = 140°$$
$$\alpha = 50° + (-1) \cdot 90° = 50° - 90° = -40°$$
$$\alpha = 50° + 2 \cdot 90° = 50° + 180° = 230°$$
$$\alpha = 50° + (-2) \cdot 90° = 50° - 180° = -130°$$

The set of RATIONAL NUMBERS is the set of all numbers which can be expressed in the form a/b with $a \in I$ and $b \in N$. We can denote the set of rational numbers by the letter Q (for quotient—we reserve R for the set of real numbers). Thus

1.5 FACTS AND FORMULAS FROM ALGEBRA

$-.3 \in Q$ since $-.3 = -3/10$, $5 \in Q$ since $5 = 5/1$, and $3.14 \in Q$ since $3.14 = 314/100$. However $\pi \notin Q$, because we cannot express π as the quotient of an integer and a natural number. Numbers which are not rational numbers are IRRATIONAL numbers, a set which can be denoted by $\bar{Q}$. In addition to π, which was already mentioned, you will find $\sqrt{5}$, $\log_{10} 3$, and $\cos 50°$ in the set of irrational numbers.

The set of REAL NUMBERS, R, is composed of the rational numbers together with the irrational numbers. A diagram of the relationship or hierarchy of these sets is shown in Figure 1–35.

Sometimes it is useful to have a notation which permits us to refer to certain sets of numbers in a brief way. One such method is SET BUILDER NOTATION. The statement

$$5 \in \{x \mid x < 10 \text{ and } x \in Q\}$$

for example, means "5 is a member of the set of numbers x such that x is less than 10 and x is a member of the set of rational numbers." The braces ({ }) indicate that we are referring to a set. The vertical bar (|) between the x's is read "such that." If the set of numbers is not specified, then it is assumed that all numbers are real numbers. We would read

$$\{(x, y) \mid y = 2x + 3\}$$

then, as "the set of ordered pairs of real numbers x and y such that y is equal to 3 more than twice the value of x," and $(0, 3)$, $(1, 5)$, $(-2, -1)$ would be several members of this set. A set denoted by $\{(x, y)\}$ would be simply a set of ordered pairs of real numbers x and y. A set denoted by $\{(x, x^2)\}$ would be a set of ordered pairs of real numbers in which the second element is the square of the first element, and $(-2, 4)$, $(0, 0)$, $(3, 9)$ would be members. Finally a set denoted by $\{(A, \tan A)\}$ would be (at this point) a set of ordered pairs of real numbers in which A would be the number of degrees in the acute angle of a right triangle and $\tan A$ would be the tangent of that angle, so that $(0, 0)$, $(45, 1)$, and $(50, \sqrt{3})$ would be members.

QUADRATIC EQUATIONS

A QUADRATIC EQUATION is an equation that can be written in the form $ax^2 + bx + c = 0$ with $a, b, c \in R$ but $a \neq 0$. These equations can always be solved by means of the quadratic formula, sometimes by factoring.

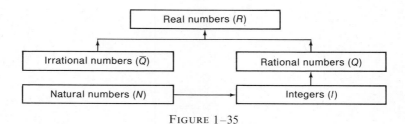

FIGURE 1–35

EXAMPLE 3   Solve the equation $x^2 - 3x + 2 = 0$ by factoring.

SOLUTION   Since $x^2 - 3x + 2 = (x-2)(x-1)$, we can write $(x-2)(x-1) = 0$. If the product of two numbers is 0, then at least one of them is 0. Thus $x - 2 = 0$ or $x - 1 = 0$, and solving these equations we obtain $x = 2$ or $x = 1$. The solution set is $\{1, 2\}$

EXAMPLE 4   Solve the equation $2x^2 + x - 6 = 0$ by factoring.

SOLUTION   The following format is fairly typical:

Equation	$2x^2 + x - 6 = 0$
Factor left member	$(2x - 3)(x + 2) = 0$
Thus	$2x - 3 = 0$ or $x + 2 = 0$
Solving these equations	$x = \dfrac{3}{2}$ or $x = -2$
Solution set	$\left\{-2, \dfrac{3}{2}\right\}$

Factoring is generally the easiest method for solving a quadratic equation, but this is not always possible, so we resort to the QUADRATIC FORMULA:

$$x = \frac{-b \pm \sqrt{b^2 - 4ac}}{2a}$$

in which a, b, and c are the values in the general quadratic equation $ax^2 + bx + c = 0$.

EXAMPLE 5   Solve the equation $2x^2 + x - 6 = 0$ by the quadratic formula.

SOLUTION   (This is the same equation that we just solved by factoring, but this is an opportunity to compare methods.) Here $a = 2, b = 1, c = -6$. Thus

$$x = \frac{-1 \pm \sqrt{(1)^2 - 4(2)(-6)}}{2(2)}$$

$$= \frac{-1 \pm \sqrt{1 + 48}}{4} = \frac{-1 \pm \sqrt{49}}{4}$$

$$= \frac{-1 \pm 7}{4}$$

1.5 FACTS AND FORMULAS FROM ALGEBRA

The double sign means $x = \dfrac{-1 + 7}{4}$ or $x = \dfrac{-1 - 7}{4}$, so $x = \dfrac{6}{4} = \dfrac{3}{2}$ or $x = \dfrac{-8}{4} = -2$. The solution set is $\left\{-2, \dfrac{3}{2}\right\}$ as before.

EXAMPLE 6   Solve the equation $2x^2 - 6x + 3 = 0$.

SOLUTION   The left member can't be factored so the quadratic formula is used. Since $a = 2$, $b = -6$, and $c = 3$,

$$x = \dfrac{-(-6) \pm \sqrt{(-6)^2 - 4(2)(3)}}{2(2)}$$

$$= \dfrac{6 \pm \sqrt{36 - 24}}{4} = \dfrac{6 \pm \sqrt{12}}{4}$$

$$= \dfrac{6 \pm 2\sqrt{3}}{4} = \dfrac{2(3 \pm \sqrt{3})}{2(2)}$$

$$= \dfrac{3 \pm \sqrt{3}}{2}$$

The solution set is thus

$$\left\{\dfrac{3 - \sqrt{3}}{2}, \dfrac{3 + \sqrt{3}}{2}\right\}$$

but if you want a decimal approximation then refer to Table V in Appendix D to find $\sqrt{3} \approx 1.732$, so then $x = (3 + \sqrt{3})/2 \approx (3 + 1.732)/2 = 4.732/2 = 2.366 \approx 2.37$, or $x = (3 - \sqrt{3})/2 \approx (3 - 1.732)/2 = 1.268/2 = 0.634$.

THE DISTANCE FORMULA

The distance between two points whose coordinates are known can be determined exactly by means of the DISTANCE FORMULA. If $P(x_1, y_1)$ and $Q(x_2, y_2)$ are the two points, then the distance PQ between them is

$$PQ = \sqrt{(x_2 - x_1)^2 + (y_2 - y_1)^2}$$

Referring to Figure 1–36, it can be seen that if lines are drawn through P and Q parallel to the x and y axes intersecting at R, a right triangle is formed and the coordinates of R are (x_2, y_1). Since $PR = x_2 - x_1$ and $QR = y_2 - y_1$, we can use the Pythagorean relation to obtain

$$(PQ)^2 = (x_2 - x_1)^2 + (y_2 - y_1)^2$$

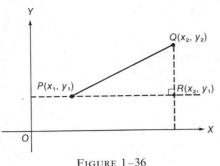

FIGURE 1–36

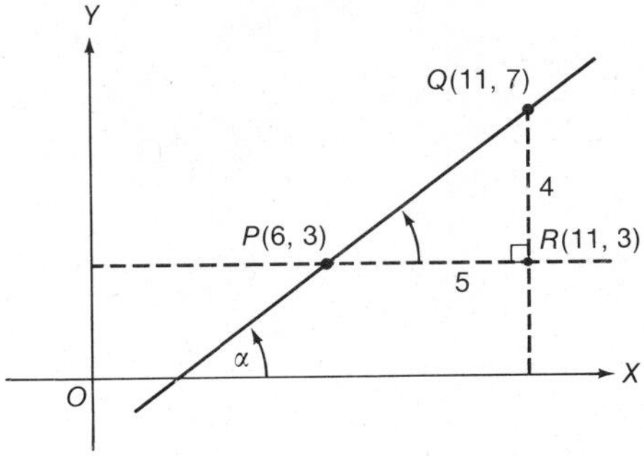

FIGURE 1–37

and taking the square root of both members,

$$PQ = \sqrt{(x_2 - x_1)^2 + (y_2 - y_1)^2}$$

SLOPE OF A LINE

The SLOPE of a line is a real number m, which is the measure of its INCLINATION, the angle the line makes with the horizontal. The slope of the line PQ of Figure 1–36 is given by

$$m = \frac{y_2 - y_1}{x_2 - x_1}.$$

In Figure 1–37 the slope of line PQ is $m = (7 - 3)/(11 - 6) = 4/5 = 0.8000$, and if α is the inclination, then angle α = angle QPR, so that $\tan \alpha = 0.8000$, and $\alpha \approx 38°40'$.

If the inclination is between $90°$ and $180°$, the slope becomes negative. In Figure 1–38, the line through $M(-2, 5)$ and $N(8, 1)$ has an inclination of more than $90°$. Calculating the slope of line MN, $m = \dfrac{1 - 5}{8 - (-2)} = \dfrac{-4}{10} = -0.4000$. The inclination can be found since angle α = angle MNU and $m°(\angle MNU) = 180° - m°(\angle \beta)$. In right triangle MTN, $\tan \beta = 4/10 = 0.4000$, so $\beta \approx 21°50'$, and $\alpha \approx 180° - 21°50' = 158°10'$.

A useful application of the slope is in the slope-intercept form of the equation of a line. If a line with slope m passes through the y axis at the point $(0, b)$ and through another point with coordinates (x, y), as in Figure 1–39, then since

$$m = \frac{y - b}{x - 0} = \frac{y - b}{x}$$

$mx = y - b$ which can be transformed into $y = mx + b$. The equation

$$y = mx + b$$

1.5 FACTS AND FORMULAS FROM ALGEBRA

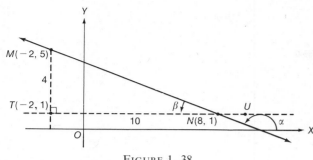

FIGURE 1–38

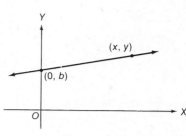

FIGURE 1–39

is the SLOPE-INTERCEPT form. From it we can read the slope, which is the coefficient of x, and the y-intercept, which is the constant b. In the equation $y = (2/3)x - 5$, for example, the slope is $2/3$ and the y-intercept is -5.

INEQUALITIES

The expression $a < b$ means "a is less than b" and $b > a$ means "b is greater than a." (The statements are equivalent.) If $a < b$, there is a positive number c such that $a + c = b$. The expression $a < x < b$ means "a is less than x and x is less than b," or "x is between a and b." For example, if $x \in \{x \mid 5 < x < 10$ and $x \in N\}$, then x can be 6, 7, 8, or 9. Although 3 is less than 10, 5 is not less than 3, so 3 is not a member of the set. Also 3 is not "between" 5 and 10. The expression $a \leq b$ means "a is less than or equal to b," and underlining the less than or greater than symbols in other situations also permits the possibility of equality. If $x \in \{x \mid 5 < x \leq 10$ and $x \in N\}$, then x can be 6, 7, 8, 9, or 10.

EXERCISES

1. List all members of the following sets:
   a. $\{x \mid x \leq 5$ and $x \in N\}$
   b. $\{x \mid -4 < x < 4$ and $x \in I\}$
   c. $x \in \{5, 22/7, 0, \sqrt{3}, -\pi, -2.1, \cos 30°\}$ and $x \in Q$.

2. List all members of the following sets:
   a. $\{x \mid -3 < x < 5$ and $x \in N\}$
   b. $\{x \mid 2 \leq x \leq 6$ and $x \in I\}$
   c. $x \in \{\sqrt{4}, 1, -3.14, \frac{\pi}{2}, -\sqrt{2}, 2\frac{3}{5}, \sin 30°\}$ and $x \in \overline{Q}$

3. Find all values of α if $\alpha = 25° + k \cdot 60°$ and $-3 \leq k \leq 3$ (Recall that $k \in I$).

4. Find all values of β if $\beta = 225° + k \cdot 180°$ and $-2 \leq k \leq 2$.

5. Find the following products:
   a. $(2x - 5)(3x + 1)$
   b. $(\tan \alpha - 2)(\tan \alpha + 2)$
   c. $(4x + 3)^2$

6. Find the following products:
   a. $(2 \sin u - 3)^2$
   b. $(5t + 2)(2t - 1)$
   c. $(x - 3y)(x + 3y)$

7. Factor each of the following expressions:
   a. $16x^2 - 25$
   b. $16x^2 - 40x + 25$
   c. $6x^2 - 13x + 5$

8. Factor each of the following expressions:
   a. $3(\sin t)^2 - 4 \sin t - 4$

b. $(\cos \alpha)^2 - 1$
c. $(\cos \alpha)^2 - 2 \cos \alpha + 1$

9. Solve by factoring:
a. $x^2 - 3x - 4 = 0$
b. $3x^2 - 10x + 3 = 0$

10. Solve by factoring:
a. $x^2 + 2x - 15 = 0$
b. $6x^2 - x - 5 = 0$

11. Solve by means of the quadratic formula (*Calculate the answers to the nearest hundredth*):
a. $x^2 - 10x + 20 = 0$
b. $9x^2 - 12x + 2 = 0$

12. Solve by means of the quadratic formula (*Calculate the answers to the nearest hundredth*):
a. $x^2 + 2x - 9 = 0$
b. $5x^2 - 4x - 2 = 0$

13. Find the distance between the following pairs of points and the slope of the line through them (*See example 7*):
a. $C(-3, 0)$ and $D(5, 6)$
b. $E(-6, 6)$ and $F(-3, 9)$

EXAMPLE 7 $A(-3, 5)$ and $B(2, 8)$

SOLUTION

$$AB = \sqrt{(x_2 - x_1)^2 + (y_2 - y_1)^2}$$
$$= \sqrt{[2 - (-3)]^2 + (8 - 5)^2}$$
$$= \sqrt{(5)^2 + (3)^2} = \sqrt{25 + 9} = \sqrt{34}$$

$$m = \frac{y_2 - y_1}{x_2 - x_1} = \frac{8 - 5}{2 - (-3)} = \frac{3}{5}$$

14. Find the distance between the following pairs of points and the slope of the line through them:
a. $G(2, -4)$ and $H(10, 11)$
b. $M(-1, -5)$ and $N(7, -9)$

15. Find the inclination of each of the following lines, correct to the nearest ten minutes:
a. Passing through $(-2, 4)$ and $(3, 9)$
b. If the slope of the line is $-3/5$

16. Find the slope of each of the following lines:
a. The equation is $3x + 2y + 5 = 0$
b. The inclination is $36°30'$

1.6 ANGLE MEASUREMENTS

An angle can be drawn in any position, as illustrated in Figure 1–40. The MEASURE of the angle is normally stated in terms of degrees, and the size of the angle depends upon the amount of opening between the sides, rather than the length of the sides. It is oftentimes desirable, however, to place the angle in what is known as standard position. An angle is in STANDARD POSITION if one side coincides with the positive ray of the x axis in a coordinate axis system and

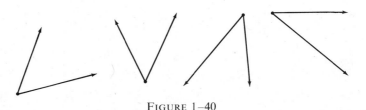

FIGURE 1–40

1.6 ANGLE MEASUREMENTS

its vertex is at the origin. The angle AOB in Figure 1–41 is in standard position. Side OB, the one which coincides with the x axis, is called the INITIAL SIDE and side OA is called the TERMINAL SIDE of the angle.

An angle can be generated by rotating the terminal side in either a clockwise or counterclockwise direction. We shall consider a counterclockwise rotation to be positive and a clockwise rotation to be negative. Thus the measure of the angle in Figure 1–42a is 40°, whereas the measure of the angle in Figure 1–42b is $-320°$.

COTERMINAL ANGLES

If the terminal sides of two angles in standard position coincide, they are said to be COTERMINAL angles. The two angles in Figure 1–42 are coterminal angles. If $m°(\angle AOB)$ is the measure in degrees of angle AOB, then all angles α such that $m°(\angle \alpha) = m°(\angle AOB) + k \cdot 360°$ (k an integer) are coterminal with angle AOB. Figure 1–43 illustrates several coterminal angles. Angles of 100°, 460°, 3700°, $-260°$, and $-620°$ are coterminal angles, since $460° = 100° + 1(360°)$, $3700° = 100° + 10(360°)$, $-260° = 100° + (-1)(360°)$, and $-620° = 100° + (-2)(360°)$. A given angle can have an infinite number of angles coterminal with it.

RADIAN MEASURE

Another useful measure of an angle is the radian. If α refers to the angle AOB in Figure 1–44, then the RADIAN MEASURE of α, or $m^R(\alpha)$, is the ratio of the length s of the arc AB intercepted on the circle to its radius r. Thus $m^R(\alpha) = \dfrac{s}{r}$.

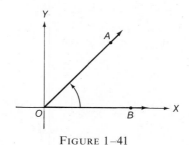

FIGURE 1–41

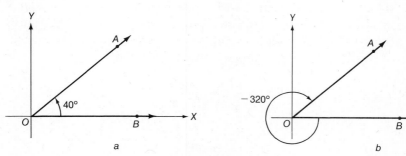

FIGURE 1–42

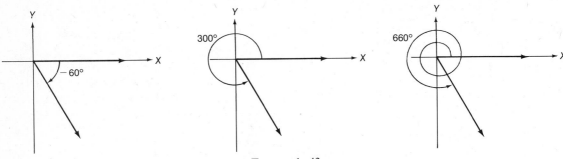

FIGURE 1-43

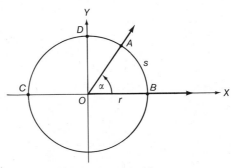

FIGURE 1-44

The key relationship used to identify angles in radians and to convert between radian and degree measure involves the circumference of a circle, $c = 2\pi r$. If r is the radius of the circle in Figure 1-44, then the length of the arc BAC, a semicircle, is πr units. Angle BOC is a straight angle which measures $180°$. Thus $m^R(\angle BOC) = (\pi r/r)^R = \pi^R$, while $m°(\angle BOC) = 180°$, so $\pi^R = 180°$. From this equation we obtain

$$1^R = \left(\frac{180}{\pi}\right)° \quad \text{and} \quad \left(\frac{\pi}{180}\right)^R = 1°$$

which enables us to change from degree to radian measure or vice versa.

Many of the problems we will be dealing with in the remainder of the text will involve angles whose measure involves some multiple of 15 degrees, so it would be worthwhile for you to be able to change these particular angles from one type of measure to the other rapidly. Note that

$$15° = 15(1°) = 15\left(\frac{\pi^R}{180}\right) = \frac{\pi^R}{12}$$

$$30° = 2(15°) = 2\left(\frac{\pi^R}{12}\right) = \frac{\pi^R}{6}$$

1.6 ANGLE MEASUREMENTS

and so on. Also,

$$\frac{5\pi^R}{6} = 5\left(\frac{\pi^R}{6}\right) = 5(30°) = 150°$$

$$\frac{\pi^R}{4} = 3\left(\frac{\pi^R}{12}\right) = 3(15°) = 45°$$

Once we begin to recognize some of the more frequently used measures, namely 30°, 45°, and 60°, the others are easy to convert, that is,

$$210° = 7(30°) = 7\left(\frac{\pi^R}{6}\right) = \frac{7\pi^R}{6}$$

$$450° = 10(45°) = 10\left(\frac{\pi^R}{4}\right) = \frac{5\pi^R}{2}$$

$$\frac{5\pi^R}{3} = 5\left(\frac{\pi^R}{3}\right) = 5(60°) = 300°$$

The rectangular coordinate axes divide the plane into four distinct QUADRANTS, as illustrated in Figure 1–45. If $m°(\alpha)$ represents the degree measure of an angle α in standard position, then if α is an angle which measures between 0° and 90° we say that α is a FIRST QUADRANT ANGLE, if α is an angle which measures between 90° and 180° then α is a SECOND QUADRANT ANGLE, and so on. If $m^R(\alpha)$ is the radian measure of an angle α in standard position, then if α is an angle which measures between 0 and $\pi/2$ radians we say that α is a first quadrant angle, while if α is an angle which measures between π and $3\pi/2$ we say that α is a third quadrant angle (since $\pi^R = 180°$, and $(3\pi/2)^R = 270°$). Actually *any* angle in standard position which has its terminal side located in the first quadrant is a first quadrant angle, so that angles of 45°, −315°, 405°, $(9\pi/4)^R$, and $-(15\pi/4)^R$ are all first quadrant angles. (All of the angles just mentioned are coterminal.) Note that $-315° = 45° + (-1)(360°)$, $405° = 45° + 1(360°)$, $405° = 405(1°) = 405(\pi/180)^R = (9\pi/4)^R$, and $-(15\pi/4)^R = (\pi/4)^R - 2(2\pi^R)$.

It is frequently convenient to refer to certain intervals of numbers that involve measures of angles. If we write $a < \alpha < b$, then α lies between the numbers a

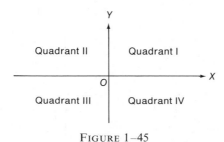

FIGURE 1–45

and b. Thus if $m°(\alpha)$ is the degree measure of an angle in standard position, then when $0 < \alpha < 90$, α is a first quadrant angle and when $270 < \alpha < 360$, α is a fourth quadrant angle. We could also say that if $0° < m°(\alpha) < 90°$, α is a first quadrant angle. Note the slight difference between the interval involving a relationship among *numbers* 0, α, and 90 as compared to a relationship among *measures* $0°$, $m°(\alpha)$, and $90°$.

We can multiply each member of an interval such as $a < \alpha < b$ by a positive number k and get an equivalent statement. If $a < \alpha < b$, then $ka < k\alpha < kb$. From time to time in the future we will want to do this. For example, if $90 < \alpha < 180$, then an equivalent interval for 3α is $270 < 3\alpha < 540$ while an equivalent interval for $(1/2)\alpha$ is $45 < (1/2)\alpha < 90$. From the latter you might conclude that if α is a second quadrant angle, $(1/2)\alpha$ is always a first quadrant angle, but that is not always the case. If $450 < \alpha < 540$ then α is a second quadrant angle, but $225 < (1/2)\alpha < 270$, so $(1/2)\alpha$ is not a first quadrant angle but rather a third quadrant angle.

We have been saying a lot about angles which terminate in one of the four different quadrants, but there are certain angles which have their terminal sides on the x or y axis, such as $90°$, $180°$, $270°$, and so on. Any angle having a measure which is a multiple of $90°$ or $(\pi/2)^R$ is a QUADRANTAL ANGLE, and is not considered to be in a quadrant.

EXERCISES

1. Draw the first angle in each of the following sets of angles in standard position; then tell which of the angles in each set are coterminal with the first.
   a. $70°$; $430°$, $-430°$, $-290°$, $790°$
   b. $-130°$; $-230°$, $230°$, $-490°$, $590°$
   c. $320°$; $680°$, $-40°$, $1040°$, $-320°$
   d. $\left(\frac{5\pi}{6}\right)^R$; $\left(\frac{17\pi}{6}\right)^R$, $\left(\frac{\pi}{6}\right)^R$, $\left(-\frac{7\pi}{6}\right)^R$, $\left(-\frac{19\pi}{6}\right)^R$

2. Draw the first angle in each of the following sets of angles in standard position; then tell which of the angles in each set are coterminal with the first.
   a. $450°$; $-270°$, $90°$, $-90°$, $810°$
   b. $-200°$; $20°$, $160°$, $-560°$, $520°$
   c. $200°$; $560°$, $160°$, $-160°$, $-520°$
   d. $\left(\frac{3\pi}{2}\right)^R$; $\left(\frac{\pi}{2}\right)^R$, $\left(-\frac{\pi}{2}\right)^R$, $\left(\frac{7\pi}{2}\right)^R$, $\left(-\frac{5\pi}{2}\right)^R$

Change each of the following to radian measure. Answers may be left in terms of π.

3. $75°$
4. $105°$
5. $120°$
6. $240°$
7. $210°$
8. $330°$
9. $90°$
10. $270°$
11. $420°$
12. $660°$
13. $-100°$
14. $-200°$
15. $155°$
16. $225°$

Change each of the following to degree measure.

17. $(7\pi/6)^R$
18. $(5\pi/6)^R$
19. $(11\pi/12)^R$
20. $(17\pi/12)^R$
21. $2\pi^R$
22. $3\pi^R$
23. $(3\pi/4)^R$
24. $(5\pi/4)^R$
25. 2^R
26. 3^R
27. $(-3\pi/2)^R$
28. $(-\pi/2)^R$

The least positive angle in a set of angles is the angle whose measure is positive and which is closest to zero. In the set $\{-160°, -20°, 70°,$

250°, 530°} the least positive angle is 70°. Find the least positive angle which is coterminal with each of the following angles.

29. 800°
30. 750°
31. −200°
32. −140°
33. $(9\pi/4)^R$
34. $(27\pi/6)^R$
35. $(-5\pi/12)^R$
36. $(-3\pi/4)^R$

Find an equivalent interval in which α lies, if

37. $0° < 2\alpha < 90°$
38. $0° < 3\alpha < 180°$
39. $0° < \alpha/2 < 90°$
40. $90° < \alpha/2 < 180°$
41. $-90° < 2\alpha < 0°$
42. $-90° < 2\alpha < 90°$
43. $180° < 2\alpha < 270°$
44. $-270° < 2\alpha < -180°$
45. $(\pi/2)^R < \alpha/2 < \pi^R$
46. $(3\pi/2)^R < 2\alpha < 2\pi^R$

Tell in what quadrant each of the following angles terminates.

47. 500°
48. −500°
49. $(19\pi/6)^R$
50. $(-7\pi/6)^R$
51. α, if $180° < 2\alpha < 270°$
52. α, if $(-2\pi)^R < 2\alpha < (-3\pi/2)^R$

REVIEW EXERCISES

In triangle ABC, angle C is 90° and sides a, b, and c are opposite angles A, B, and C, respectively.

1. Find side b if angle $A = 37°$ and $c = 30$.

2. Find side b if angle $A = 53°$ and $a = 120$.

3. Find angle A if side $a = 20$ and side $b = 50$ (Nearest ten minutes).

4. At a point 100 feet from the foot of a flagpole, the angle of elevation to the top of the pole is 31°. Find the height of the pole to the nearest foot.

5. Five holes are to be equally spaced, each 5.00 inches from the center of a circular metal disc. What is the distance between centers of two adjacent holes?

6. The power factor (PF) of an electric motor is determined by $PF = \cos A$, in which the angle A may be found by the formula

$$\tan A = \frac{X_L}{R}$$

where X_L is the inductive reactance for the circuit and R is its resistance. If $X_L = 30$ and $R = 20$, find PF.

7. Find the area of a triangle whose sides of 8 inches and 12 inches include an angle of 42°.

8. A plane is flying at an altitude of 10,000 feet when its navigator sights two objects on the ground directly ahead. What is the distance from object A to object B if the angle of depression of A is 32° when the angle of depression of B is 26°? (Refer to Figure 1–46)

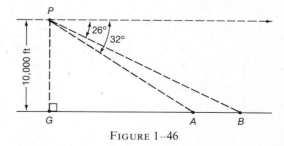

FIGURE 1–46

9. The sector AOB of the circle in Figure

1–47 is the set of points bounded by angle AOB and arc AB of the circle. It is a fact that the ratio of the area of the sector to the area of the circle is equal to the ratio of the length of arc AB to the circumference of the circle. Show that if θ is the measure of angle AOB in radians, the area of the sector is given by $S = \frac{1}{2}r^2\theta$. (Recall that $\theta = s/r$.)

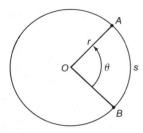

FIGURE 1–47

10. Find the area of the sector of a circle which has radius 20 inches if the central angle of the sector is 50°.

11. The coordinates of two points on a circle which is centered at the origin 0 are $P(-12, 5)$ and $Q(5, 12)$.
   a. Find the distance PQ.
   b. If R is the point at which the circle intersects the x axis on the positive side, what are the coordinates of R?
   c. What is the distance from Q to R?
   d. Find the measure of angle POQ.
   e. Find the measure of angle QOR.

12. If θ is an acute angle such that $u = \sin \theta$ and $3u^2 + 2u - 1 = 0$, find the value of θ, correct to the nearest 10 minutes.

13. If α is an acute angle such that $t = \cos \alpha$ and $20t^2 - 20t + 3 = 0$, find the greater value of α, correct to the nearest 10 minutes.

14. Find the area of the segment of the circle in Figure 1–48 (region bounded by line segment XY and arc XY) if $m(\angle XOY) = 4\pi/9$ radians and radius $OX = 15$ cm.

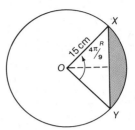

FIGURE 1–48

15. If the slope of one line is $3/5$ and the slope of a second line is $12/5$, find the smaller angle formed when they intersect.

16. According to Snell's Law, the law of refraction, the index of refraction for light waves traveling through air which strike a surface such as glass at an angle θ_1, then travel through the glass at angle θ_2 (Figure 1–49) is given by

$$n = \frac{\sin \theta_1}{\sin \theta_2}$$

   a. If $\theta_1 = 32°$ and $\theta_2 = 27°$, find the index of refraction.
   b. If $n = 1.5$ and $\theta_2 = 15°$, find θ_1.

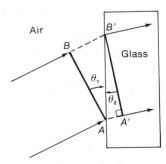

FIGURE 1–49

CHAPTER TWO

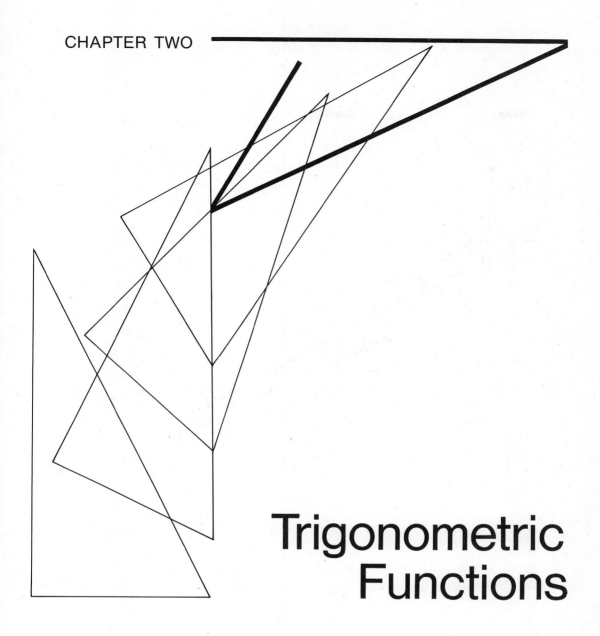

Trigonometric Functions

2.1 THE TRIGONOMETRIC RATIOS; FUNDAMENTAL IDENTITIES

In the previous chapter we defined the trigonometric ratios in terms of the sides of a right triangle. Now we are ready to generalize our definitions so that they are more widely applicable and useful, as well as to define two additional trigonometric ratios, SECANT and COSECANT, for which the abbreviations are sec and csc.

Assume that α is an angle in standard position and that P is a point on the terminal side of α which is at a positive distance r from the origin (Refer to Figure 2–1).

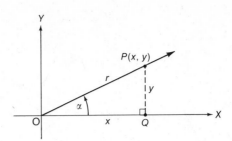

FIGURE 2–1

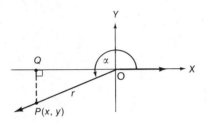

FIGURE 2–2

If the coordinates of P are (x, y), then $r = \sqrt{x^2 + y^2}$, where $r \neq 0$, and we define

$$\sin \alpha = \frac{y}{r} \qquad \csc \alpha = \frac{r}{y} \, (y \neq 0)$$

$$\cos \alpha = \frac{x}{r} \qquad \sec \alpha = \frac{r}{x} \, (x \neq 0)$$

$$\tan \alpha = \frac{y}{x} \, (x \neq 0) \qquad \cot \alpha = \frac{x}{y} \, (y \neq 0)$$

Note that $\sin \alpha$ and $\cos \alpha$ are always defined, but there are certain instances when the other ratios are not. Note also that if α is a positive acute angle and if PQ is drawn perpendicular to the x axis, then POQ is a right triangle and that the definitions given in Chapter One still hold.

The terminal side of α can be in any quadrant or can be a quadrantal angle, as indicated in Figure 2–2. If we know any two of the three values x, y, and r, then we can calculate the third and find all six ratios.

EXAMPLE 1   Given that α terminates in Quadrant II and that the terminal side of α passes through $(-5, 12)$, find all six trigonometric ratios of α.

SOLUTION   It is often helpful to draw a sketch of α, as in Figure 2–3. Then

$$r = \sqrt{(-5)^2 + (12)^2} = \sqrt{25 + 144} = \sqrt{169} = 13$$

2.1 THE TRIGONOMETRIC RATIOS; FUNDAMENTAL IDENTITIES

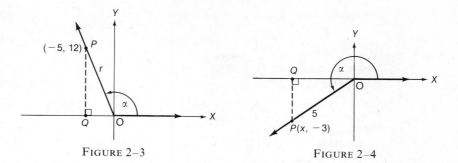

FIGURE 2-3   FIGURE 2-4

and by the definition

$$\sin \alpha = \frac{12}{13} \qquad \csc \alpha = \frac{13}{12}$$

$$\cos \alpha = \frac{-5}{13} \qquad \sec \alpha = \frac{-13}{5}$$

$$\tan \alpha = \frac{-12}{5} \qquad \cot \alpha = \frac{-5}{12}$$

EXAMPLE 2 Given that $\sin \alpha = -3/5$ and that α terminates in Quadrant III, find the other five trigonometric ratios of α.

SOLUTION   Although $\sin \alpha = y/r$ by definition, this does not necessarily imply that if $y/r = -3/5$, $y = -3$ and $r = 5$. We could have $y/r = -3/5 = -6/10 = -1.5/2.5$, and so on. If we do choose to let $r = 5$ in this case, however, we can draw Figure 2-4 and complete the solution. Since $r = \sqrt{x^2 + y^2}$, $r^2 = x^2 + y^2$. Thus $5^2 = x^2 + (-3)^2$, or $x^2 = 25 - 9 = 16$, so $x = 4$ or $x = -4$. We choose $x = -4$ since α terminates in Quadrant III. Then

$$\sin \alpha = \frac{-3}{5} = -\frac{3}{5} \qquad \csc \alpha = \frac{5}{-3} = -\frac{5}{3}$$

$$\cos \alpha = \frac{-4}{5} = -\frac{4}{5} \qquad \sec \alpha = \frac{5}{-4} = -\frac{5}{4}$$

$$\tan \alpha = \frac{-3}{-4} = \frac{3}{4} \qquad \cot \alpha = \frac{-4}{-3} = \frac{4}{3}$$

Note that had we chosen $r = 10$, $y = -6$, the ratios would be the same because the right triangles involved would be similar, as demonstrated in Section 1.2.

Another feature observable from the definitions and from the examples is the reciprocal relation between certain pairs of ratios, namely sine and cosecant, cosine and secant, tangent and cotangent. In other words, in each case $\sin \alpha \csc \alpha = 1$, since $\left(\dfrac{y}{r}\right)\left(\dfrac{r}{y}\right) = 1$, and likewise $\cos \alpha \sec \alpha = 1$, $\tan \alpha \cot \alpha = 1$. This makes it easier to remember the definitions and calculate specific values, since $\csc \alpha = 1/\sin \alpha$, $\sec \alpha = 1/\cos \alpha$, and $\cot \alpha = 1/\tan \alpha$ so long as the denominators are not zero.

EXAMPLE 3   Find the other five trigonometric ratios if $0° < \alpha < 90°$ and $\cos \alpha = \tfrac{1}{2}$.

SOLUTION   Draw Figure 2–5. Again assume $r = 2$ and $x = 1$, since $\cos \alpha = x/r$. Then $2^2 = 1^2 + y^2$ or $y^2 = 3$, so that $y = \pm\sqrt{3}$. Choose $y = \sqrt{3}$ since α is in the first quadrant. Then

$$\sin \alpha = \frac{\sqrt{3}}{2} \qquad \csc \alpha = \frac{2}{\sqrt{3}} = \frac{2}{\sqrt{3}} \cdot \frac{\sqrt{3}}{\sqrt{3}} = \frac{2\sqrt{3}}{3}$$

$$\cos \alpha = \frac{1}{2} \qquad \sec \alpha = 2$$

$$\tan \alpha = \sqrt{3} \qquad \cot \alpha = \frac{1}{\sqrt{3}} = \frac{1}{3}\sqrt{3}$$

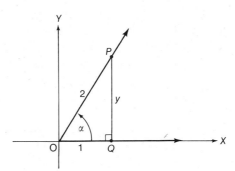

FIGURE 2–5

The algebraic signs of the ratios in each quadrant are important to know and to recognize. Since the value of r is always positive, knowing the definitions of the trigonometric ratios and the signs of the abscissa x and the ordinate y in each quadrant is sufficient to determine the sign of the ratio. In the first quadrant both x and y are positive, so all of the ratios are positive. In the second quadrant x is negative and y is positive so $\sin \alpha = y/r$ is positive, as is its reciprocal $\csc \alpha$, and $\cos \alpha = x/r$ is negative along with its reciprocal $\sec \alpha$, while $\tan \alpha = y/x$ is negative and $\cot \alpha = x/y$ is negative. Each pair of reciprocal ratios must have the same sign, since their product is 1. Thus we can complete Table 2–1. The ASTC at the foot of the table can be used as a mnemonic device for remembering it,

2.1 THE TRIGONOMETRIC RATIOS; FUNDAMENTAL IDENTITIES

TABLE 2–1

		Quadrants			
		I	II	III	IV
Ratios	sin, csc	+	+	−	−
	cos, sec	+	−	−	+
	tan, cot	+	−	+	−
		A	S	T	C

referring to All, Sine, Tangent, Cosine (and the reciprocals, of course) positive, while the rest are negative. Some say "All Scholars Take Calculus" as an aid in remembering the order of the letters.

EXAMPLE 4 In what quadrant does α terminate if $\sin \alpha < 0$ and $\tan \alpha > 0$?

SOLUTION Since $\sin \alpha < 0$, α must terminate in Quadrants III or IV. If $\tan \alpha > 0$, then α must terminate in Quadrants I or III. Thus α must terminate in Quadrant III.

EXAMPLE 5 If $\cot \alpha = -24/7$ and $\cos \alpha > 0$, find the values of the remaining five trigonometric ratios.

SOLUTION Since $\cos \alpha > 0$ in Quadrants I and IV only, and all ratios are positive in Quadrant I, α must be in Quadrant IV. Draw Figure 2–6. Let $x = 24$ and $y = -7$, and calculate $r^2 = (24)^2 + (-7)^2 = 576 + 49 = 625$. Then $r = 25$, and

$$\sin \alpha = -\frac{7}{25} \qquad \csc \alpha = -\frac{25}{7}$$

$$\cos \alpha = \frac{24}{25} \qquad \sec \alpha = \frac{25}{24}$$

$$\tan \alpha = -\frac{7}{24} \qquad \cot \alpha = -\frac{24}{7}$$

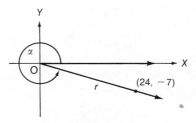

FIGURE 2–6

EXERCISES

Given that the terminal side of α passes through the given point, find all six trigonometric ratios for α. Sketch the angle.

1. $(-6, 8)$
2. $(-8, -6)$
3. $(2, -2)$
4. $(-\sqrt{3}, 1)$
5. $(-12, -5)$
6. $(-7, 24)$

Given that the terminal side of α lies in Quadrant I, find the other five trigonometric ratios.

7. $\tan \alpha = 3/4$
8. $\cot \alpha = 12/5$
9. $\sin \alpha = \sqrt{11}/6$
10. $\cos \alpha = \sqrt{5}/3$
11. $\sec \alpha = 17/15$
12. $\csc \alpha = 25/24$

Determine the other five trigonometric ratios for α.

13. $\sin \alpha = -4/5$ and $\cos \alpha > 0$
14. $\cos \alpha = -12/13$ and $\sin \alpha > 0$
15. $\tan \alpha = \sqrt{23}/11$ and $\sin \alpha < 0$
16. $\csc \alpha = 11/\sqrt{21}$ and $\tan \alpha < 0$

Determine all six trigonometric ratios for each of the following angles. Hint: Sketch the angle, let $r = 2$.

17. $150°$
18. $225°$
19. $180°$
20. $270°$

21. Refer to the examples of this section and some of the exercises you have worked previously, and in each case divide the value of $\sin \alpha$ by the value of $\cos \alpha$ and compare the result with $\tan \alpha$. You should be able to reach a conclusion, then prove it by means of the definitions of the trigonometric ratios.

22. Refer to the examples of this section and some of the exercises you have worked previously, and in each case find the square of $\sin \alpha$ and the square of $\cos \alpha$, then add the two squares. You should be able to reach a conclusion, then prove it by means of the definitions.

2.2 SPECIAL ANGLES

The trigonometric ratios for angles in multiples of 10 minutes and also of 0.01 radian are given in Tables I and II in Appendix D. More extensive tables exist, and there are formulas which enable them to be found nearly instantaneously by means of computers. There are certain angles, however, which appear so frequently in mathematical literature (including the remainder of this text), that the values of their trigonometric ratios should be memorized or at least be available rapidly by means of a quick sketch. These angles are all multiples of 30 or 45 degrees—$0°$, $30°$, $45°$, $60°$, $90°$, $120°$, $135°$, $150°$, $180°$, to name a few. Our technique will be to locate a point on the terminal side at an arbitrary distance from the origin, then use the basic definitions, as in the previous section.

EXAMPLE 1 Find the trigonometric ratios for an angle of $0°$.

SOLUTION Let $r = 1$, and draw a sketch as in Figure 2–7. Note that $x = 1$, $y = 0$. Then

$$\sin 0° = \frac{0}{1} = 0 \qquad \csc 0° \text{ is undefined}$$

2.2 SPECIAL ANGLES

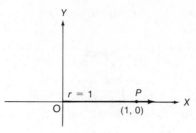

FIGURE 2–7

$$\cos 0° = \frac{1}{1} = 1 \qquad \sec 0° = \frac{1}{1} = 1$$

$$\tan 0° = \frac{0}{1} = 0 \qquad \cot 0° \text{ is undefined}$$

EXAMPLE 2   Find the trigonometric ratios for an angle of 270°.

SOLUTION   Let $r = 1$, and draw Figure 2–8. Note that $x = 0$, $y = -1$. Then

$$\sin 270° = \frac{-1}{1} = -1 \qquad \csc 270° = \frac{1}{-1} = -1$$

$$\cos 270° = \frac{0}{1} = 0 \qquad \sec 270° \text{ is undefined}$$

$$\tan 270° \text{ is undefined} \qquad \cot 270° = \frac{0}{-1} = 0$$

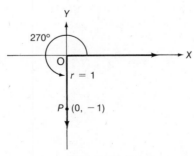

FIGURE 2–8

You will find that the values of sine and cosine for all quadrantal angles are either $-1, 0$, or 1; the tangent and cotangent of these angles are either 0 or undefined; and secant and cosecant are $-1, 0, 1$ or undefined. If you merely think of the point on the appropriate axis with its coordinates, the values are easy to work out mentally.

EXAMPLE 3   Find the trigonometric ratios for a 30° angle.

SOLUTION   Let $r = 1$, and draw a sketch as in Figure 2–9. Again recall that in a 30°, 60°, right triangle the side opposite the 30° angle is half the length of the hypotenuse. Thus $y = PQ = 1/2$, and since $x^2 + y^2 = r^2$, $x^2 + 1/2^2 = 1^2$, or $x^2 = 1 - 1/4 = 3/4$, so that $x = \pm\sqrt{3}/2$. Since P is in Quadrant I we choose $x = +\sqrt{3}/2 = 1/2\sqrt{3}$. Then

$$\sin 30° = \frac{1/2}{1} = \frac{1}{2} \qquad \csc 30° = \frac{1}{1/2} = 2$$

$$\cos 30° = \frac{1/2\sqrt{3}}{1} = \frac{1}{2}\sqrt{3} \qquad \sec 30° = \frac{1}{1/2\sqrt{3}} = \frac{2}{\sqrt{3}}$$

$$\tan 30° = \frac{1/2}{1/2\sqrt{3}} = \frac{1}{\sqrt{3}} \qquad \cot 30° = \frac{1/2\sqrt{3}}{1/2} = \sqrt{3}$$

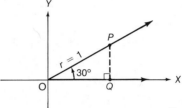

FIGURE 2–9

EXAMPLE 4   Find the trigonometric ratios for a 150° angle.

SOLUTION   This time let $r = 2$. Draw Figure 2–10. We find that in this case $y = 1$, $x = -\sqrt{3}$, so the ratios are

$$\sin 150° = \frac{1}{2} \qquad \csc 150° = \frac{2}{1} = 2$$

$$\cos 150° = \frac{-\sqrt{3}}{2} = -\frac{1}{2}\sqrt{3} \qquad \sec 150° = \frac{2}{-\sqrt{3}} = -\frac{2}{\sqrt{3}}$$

$$\tan 150° = \frac{1}{-\sqrt{3}} = -\frac{1}{\sqrt{3}} \qquad \cot 150° = \frac{-\sqrt{3}}{1} = -\sqrt{3}$$

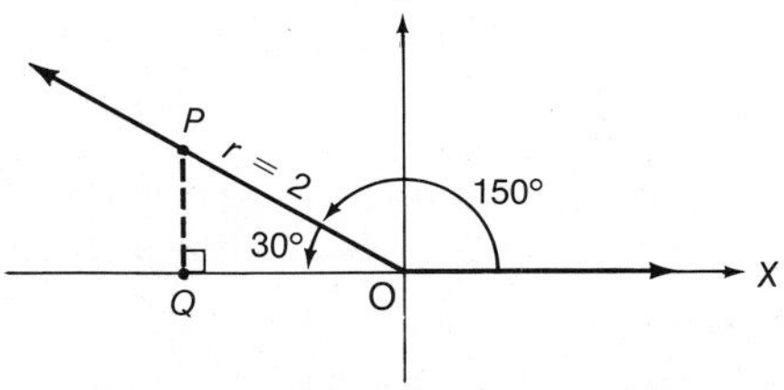

FIGURE 2–10

2.2 SPECIAL ANGLES

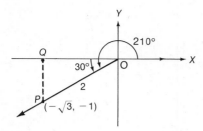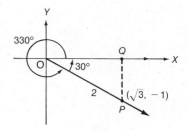

FIGURE 2–11

Note that the above values are exactly the same as those for 30° except for signs. It can be seen that (except for signs) angles of 210° and 330° would have the same values as 30° and 150°. Refer to Figure 2–11.

EXAMPLE 5   Find the trigonometric ratios for a 60° angle.

SOLUTION   Let $r = 2$ and draw Figure 2–12. This time the x and y values are changed, but the general relationships remain the same since this is, after all, a 30°, 60°, right triangle. Thus

$$\sin 60° = \frac{\sqrt{3}}{2} = \frac{1}{2}\sqrt{3} \qquad \csc 60° = \frac{2}{\sqrt{3}}$$

$$\cos 60° = \frac{1}{2} \qquad \sec 60° = \frac{2}{1} = 2$$

$$\tan 60° = \frac{\frac{1}{2}\sqrt{3}}{\frac{1}{2}} = \sqrt{3} \qquad \cot 60° = \frac{\frac{1}{2}}{\frac{1}{2}\sqrt{3}} = \frac{1}{\sqrt{3}}$$

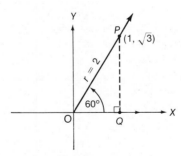

FIGURE 2–12

You will be asked to verify that the ratios for 120°, 240°, and 300° have the same values as those for 60°, except for sign. Recall the signs of the trigonometric ratios for angles in the different quadrants, then see if you can write out the values for 240° without referring to a sketch.

EXAMPLE 6   Find the trigonometric ratios for a 45° angle.

SOLUTION   Let $r = 1$ and draw a sketch as in Figure 2–13. The triangle is isosceles, so $x = y$ and $x^2 + x^2 = 1^2$, or $2x^2 = 1$, $x^2 = 1/2$, and $x = \pm 1/\sqrt{2}$. Since the angle is in Quadrant I, we choose $x = 1/\sqrt{2}$, and it follows that $y = 1/\sqrt{2}$. From the definitions we obtain

$$\sin 45° = \frac{1}{\sqrt{2}} = \tfrac{1}{2}\sqrt{2}* \qquad \csc 45° = \sqrt{2}$$

$$\cos 45° = \frac{1}{\sqrt{2}} = \tfrac{1}{2}\sqrt{2} \qquad \sec 45° = \sqrt{2}$$

$$\tan 45° = 1 \qquad \cot 45° = 1$$

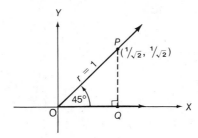

FIGURE 2–13

It can be shown that the ratios for 135°, 225°, and 315° have the same values as those of 45°, except for sign.

We are now ready to summarize the results of some of our examples in Table 2–2. It should be possible for you to reproduce the table from memory in about 60 seconds after a little practice. At first it appears to be quite a job of memorization, but after examining it you will see that the cofunctions (sine and cosine, tangent and cotangent, secant and cosecant) have the same values except in reverse order. The reciprocals have already been discussed, but checking the table will reaffirm what was proved in Section 2.1, which is the reason for leaving $\sin 45°$ in the form $1/\sqrt{2}$ rather than changing it to the equivalent $\tfrac{1}{2}\sqrt{2}$. This would seem to reduce the task to memorizing only sine and tangent (since cosine values are in reverse order of sine values, while cotangent, secant, and cosecant are reciprocals), but there is another relationship which simplifies it even more. In problem 21 of the previous section you may have observed and proved that $\tan \alpha = \sin \alpha / \cos \alpha$, since $y/x = \dfrac{y/r}{x/r}$. Look at Table 2–2 again and you will see that this is true for every angle.

All that is necessary, then, is to know the values of the sine ratios for these angles, then

1   write the cosine values in the reverse order.

* Remember that $\dfrac{1}{\sqrt{2}} = \dfrac{1}{\sqrt{2}} \cdot \dfrac{\sqrt{2}}{\sqrt{2}} = \dfrac{\sqrt{2}}{2} = \dfrac{1}{2}\sqrt{2}$.

2.2 SPECIAL ANGLES

TABLE 2–2

α	0°	30°	45°	60°	90°
sin α	0	$\dfrac{1}{2}$	$\dfrac{1}{\sqrt{2}}$	$\dfrac{\sqrt{3}}{2}$	1
cos α	1	$\dfrac{\sqrt{3}}{2}$	$\dfrac{1}{\sqrt{2}}$	$\dfrac{1}{2}$	0
tan α	0	$\dfrac{1}{\sqrt{3}}$	1	$\sqrt{3}$	—
cot α	—	$\sqrt{3}$	1	$\dfrac{1}{\sqrt{3}}$	0
sec α	1	$\dfrac{2}{\sqrt{3}}$	$\sqrt{2}$	2	—
csc α	—	2	$\sqrt{2}$	$\dfrac{2}{\sqrt{3}}$	1

2   find the tangents by dividing sin α by cos α.

3   find the cotangents, secants, and cosecants by means of reciprocals.

Now if only one could remember the sine values! You will probably have them memorized already, but if not or if you should ever forget, you can find these values by taking one-half the square roots of 0, 1, 2, 3, and 4: $\frac{1}{2}\sqrt{0} = \frac{1}{2} \cdot 0 = 0$, $\frac{1}{2}\sqrt{1} = \frac{1}{2} \cdot 1 = \frac{1}{2}$, $\frac{1}{2}\sqrt{2}$, $\frac{1}{2}\sqrt{3}$, and $\frac{1}{2}\sqrt{4} = \frac{1}{2} \cdot 2 = 1$.

EXERCISES

Draw each of the following angles in standard position, and locate a point P on the terminal side 1 or 2 units from the origin. Determine the coordinates of P, then find the trigonometric ratios for the given angles. (Try to write the ratios before doing any of the above, then check with the drawing.)

1. 90°
2. 180°
3. 210°
4. 330°
5. 120°
6. 240°
7. 300°
8. 135°
9. 225°
10. 315°

11. Prepare a table like Table 2–2 and list the values of the trigonometric ratios for the quadrantal angles 0°, 90°, 180°, 270°, 360°.

12. Prepare a table like Table 2–2 except this time write each angle in radian measure.

Find the exact value of each of the following. Note: The expression $\sin^2 \alpha$ means $(\sin \alpha)^2$. It is used rather than $\sin \alpha^2$ in order to avoid confusion with the possible interpretation $\sin \alpha^2 = \sin (\alpha)^2$.

13. sin 30° cos 60° + cos 30° sin 60°
14. cos 60° cos 30° − sin 60° sin 30°
15. 2 sin 45° cos 45°
16. $2 \cos^2 45° - 1$
17. $\sin^2 60° + \cos^2 60°$
18. $\sec^2 \pi/3 - \tan^2 \pi/3$
19. $\sin^2 45° + \cos^2 45°$

20. $\csc^2 \pi/4 - \cot^2 \pi/4$
21. $\cos^2 30° - \sin^2 30°$
22. $1 - 2\sin^2 30°$
23. $\cos^2 60° - \sin^2 60°$
24. $\cos^2 120° - \sin^2 120°$
25. $\dfrac{\sin 60° \cdot \sin 30°}{\tan 60°}$
26. $\dfrac{(\cos \pi/4)(\sin \pi/4)}{\sec \pi/3}$

Verify that each of the following statements is true:

27. $\cos 90° = \cos^2 45° - \sin^2 45°$
28. $\tan 180° = \dfrac{\tan 120° + \tan 60°}{1 - \tan 120° \tan 60°}$
29. $\sin 60° \cos 30° = \tfrac{1}{2}(\sin 90° + \sin 30°)$
30. $\cos 120° \cos 60° = \tfrac{1}{2}(\cos 180° + \cos 60°)$
31. $\sin 120° + \sin 60° = 2 \sin 90° \cos 30°$
32. $\cos 30° - \cos 90° = 2 \sin 60° \cos 30°$

2.3 FUNCTIONS OF ANY ANGLE

In the previous section we found that, except for signs, angles of 30°, 150°, 210°, and 330° all had the same numerical values for their sine, cosine, and tangent ratios. Tables I and II in Appendix D give values of trigonometric ratios only for angles of 0° through 90° or 0 through $\pi/2$ radians. Since it is sometimes necessary to find ratios for larger angles, such as 130° or 1300°, for example, it would be desirable to have a way to extend the tables. If we consider the two triangles in Figure 2–14, we see that the angle we call α_3 (since it terminates in Quadrant III) measures 180° more than α, which is a positive acute angle, since *POR* is a straight angle. We also note that angle *SOR* has the same measure as angle *POQ* since they are vertical angles. If $OP = OR = r$ and if *PQ* and *RS* are both perpendicular to the *x* axis, then triangle *SOR* is congruent to triangle *POQ* by a theorem from geometry, since they are right triangles having their hypotenuses and an acute angle, respectively, equal in measure. This means that $OQ = OS$ and that $RS = PQ$ since they are corresponding parts of congruent triangles. The sides of a triangle are always considered to be of positive length, so if the coordinates of point *R* are (x, y), we see that the coordinates of *P* are $(|x|, |y|)$. Thus $\sin \alpha_3 = y/r$ and since $\sin \alpha = |y|/r$ we can say that $\sin \alpha = |\sin \alpha_3|$. Likewise $\cos \alpha_3 = x/r$, $\cos \alpha = |x|/r$ so that $\cos \alpha = |\cos \alpha_3|$, and $\tan \alpha_3 = y/x$, $\tan \alpha = |y|/|x|$ so that $\tan \alpha = |\tan \alpha_3|$. The ratios are the same except for the algebraic sign, and all we need to do is relate the angles α_3 and α.

In order to do this we define the REFERENCE ANGLE of an angle in standard position to be the positive acute angle whose sides are the *x* axis and the terminal side of the given angle. In Figure 2–15 we see the reference angles for several different angles. (Note that the reference angle as we define it always involves the *x* and not the *y* axis.) The symbol for the reference angle will be an underlined alpha, $\underline{\alpha}$.

2.3 FUNCTIONS OF ANY ANGLE

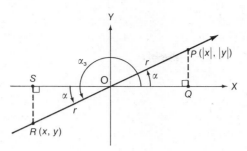

FIGURE 2–14

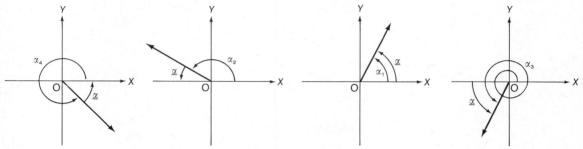

FIGURE 2–15

EXAMPLE 1 Sketch each of the following, then determine their reference angles.   *a.* 200°   *b.* 320°   *c.* 780°

SOLUTION (See Figure 2–16.)   *a.* $\underline{\alpha} = 200° - 180° = 20°$
b. $\underline{\alpha} = 360° - 320° = 40°$ *c.* $\underline{\alpha} = 780° - 2\cdot360° = 60°$

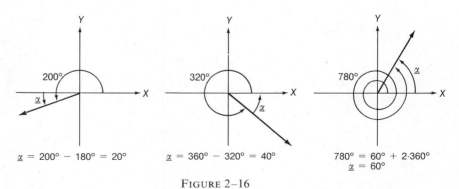

FIGURE 2–16

As shown above, if α_2 and α_4 are second and fourth quadrant angles with a reference angle $\underline{\alpha}$, then

$$|\sin \alpha_2| = |\sin \alpha_4| = \sin \underline{\alpha}$$

$$|\cos \alpha_2| = |\cos \alpha_4| = \cos \underline{\alpha}$$

$$|\tan \alpha_2| = |\tan \alpha_4| = \tan \underline{\alpha}$$

and likewise for the ratios cotangent, secant, and cosecant. Since $\underline{\alpha}$ is a positive acute angle, Tables I and II in Appendix D permit us to find the value of the trigonometric ratio for any angle.

EXAMPLE 2   Sketch each of the following angles, determine the reference angle, then find the trigonometric ratio indicated. Use $\pi \approx 3.14$ where appropriate.   *a.* sin 310°   *b.* tan 215°30′   *c.* cos 1.72^R *d.* cot 5.00^R *e.* cos 1360°

SOLUTION   (See Figure 2–17.)   *a.* $\underline{\alpha} = 360° - 310° = 50°$, $|\sin 310°| =$ sin 50° = 0.7660 from Table I in Appendix D. Since 310° terminates in

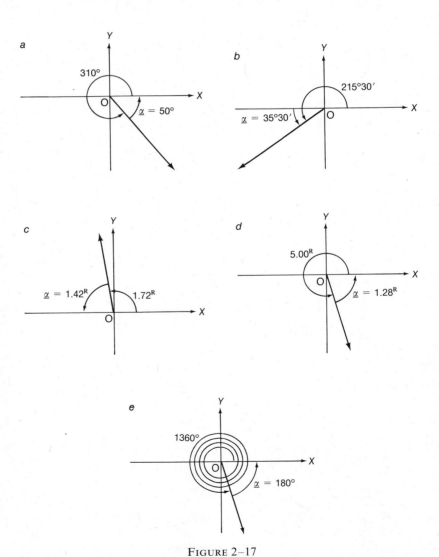

FIGURE 2–17

2.3 FUNCTIONS OF ANY ANGLE

Quadrant IV, sin 310° = −0.7660. *b.* $\underline{\alpha} = 215°30' - 180° = 35°30'$, |tan 215°30'| = tan 35°30' = 0.7133 by Table I in Appendix D. Since 215°30' terminates in Quadrant III, tan 215°30' = 0.7133. *c.* $\underline{\alpha} = (3.14 - 1.72)^R = 1.42^R$, |cos 1.72^R| = cos 1.42^R = 0.1502 by Table II in Appendix D. Since 1.72^R terminates in Quadrant II, cos 1.72^R = −0.1502. Now is perhaps a good time to observe that since $\pi/2 \approx 1.57$, $\pi \approx 3.14$, $3\pi/2 \approx 4.71$, and $2\pi \approx 6.28$, $0 < \alpha_1 < 1.57$, $1.57 < \alpha_2 < 3.14$, $3.14 < \alpha_3 < 4.17$, and $4.17 < \alpha_4 < 6.28$ for angles in the respective quadrants. *d.* $\underline{\alpha} = (6.28 - 5.00)^R = 1.28^R$, |cot 5.00^R| = cot 1.28^R = 0.2993. Since 5.00^R terminates in Quadrant IV, cot 5.00^R = −0.2993. *e.* 1360° = 280° + 3·360°, so $\underline{\alpha} = 360° - 280° = 80°$. Since 1360° and 280° are coterminal angles, their trigonometric ratios must be equal. Thus |cos 1360°| = |cos 280°| = cos 80° = 0.1736. Since 1360° terminates in Quadrant IV, cos 1360° = 0.1736.

Thus far we have concerned ourselves exclusively with finding the trigonometric ratios when we are given an angle, but the related problem of finding an angle, given a trigonometric ratio, is just as important.

EXAMPLE 3 Find α if tan α = 0.5467.

SOLUTION In Table I in Appendix D we find that tan 28°40' = 0.5467. If α is an angle of a right triangle then α = 28°40'. On the other hand, $\underline{\alpha}$ might be the reference angle for any number of angles α_1 or α_3 which terminate in Quadrant I or Quadrant III, since tangent is positive in these quadrants. If we make a sketch such as that in Figure 2–18, we see that in the first quadrant $\alpha_1 = 28°40' + k \cdot 360°$ and in the third quadrant $\alpha_3 = 208°40' + k \cdot 360°$.

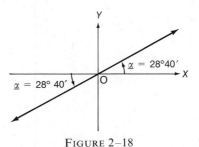

FIGURE 2–18

EXAMPLE 4 Find the least positive angle for which cos α = −0.6180.

SOLUTION Since cos $\alpha < 0$, α must terminate in Quadrants II or III. We choose α_2, however, because $90° < \alpha_2 < 180°$ whereas $180° < \alpha_3 < 270°$. Now cos $\underline{\alpha}$ = 0.6180 if $\underline{\alpha}$ = 51°50', and α_2 = 180° − 51°50' = 128°10'. See Figure 2–19 on the next page.

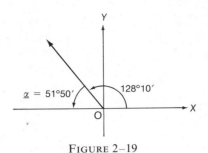

FIGURE 2-19

EXAMPLE 5   Find all angles α such that $\sin \alpha = -0.8843$ and $0° < \alpha < 360°$.

SOLUTION   Since $\sin \alpha < 0$, α must terminate in Quadrants III or IV. Now since $\sin \alpha = 0.8843$ if $\alpha = 62°10'$, $\alpha_3 = 180° + 62°10' = 242°10'$ and $\alpha_4 = 360° - 62°10' = 297°50'$.

EXAMPLE 6   Find all angles α such that $\tan \alpha = -1.210$. Use radian measure and $\pi \approx 3.14$.

SOLUTION   Since $\tan \alpha < 0$, α must terminate in Quadrants II or IV. From Table II in Appendix D we find that $\tan 0.88^R = 1.210$, so $\alpha_2 = (3.14 - 0.88)^R + k \cdot 2\pi = 2.26 + k \cdot 2\pi$ and $\alpha_4 = (6.28 - 0.88)^R + k \cdot 2\pi = 5.40 + k \cdot 2\pi$, or simply $\alpha = 2.26 + k \cdot \pi$.

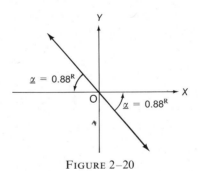

FIGURE 2-20

EXAMPLE 7   Find all angles α such that $\cos \alpha = 0.3090$ and $-\pi/2 < \alpha < \pi/2$.

SOLUTION   Since $\cos \alpha > 0$, α must terminate in Quadrants I or IV. Consulting Table II in Appendix D we find that $\cos 1.26^R = 0.3058$ and $\cos 1.25^R = 0.3153$. Since the difference between the given value, 0.3090, is 0.0032 for $\cos 1.26^R$ and 0.0063 for $\cos 1.25^R$, we choose $\alpha = 1.26^R$. Then $\alpha_1 = 1.26^R$ and $\alpha_4 = -1.26^R$.

EXERCISES

Sketch each of the following angles, then determine their reference angles.

1. 140°
2. 230°
3. 280°
4. 340°
5. 840°
6. 1040°
7. −210°
8. −510°

Sketch each of the following angles, determine the reference angles, then find the trigonometric ratio indicated. Use $\pi \approx 3.14$ where appropriate.

9. cot 260°
10. sec 320°
11. cos 110°
12. tan 212°
13. sin 410°
14. tan 532°
15. cot (−395°)
16. sec (−452°)
17. cos 1.60^R
18. sin 3.54^R
19. tan 8.28^R
20. cot (−9.00^R)

Find all angles α such that $0° \leq \alpha < 360°$ and

21. sin $\alpha = 0.5125$
22. tan $\alpha = 2.414$
23. cot $\alpha = -0.6009$
24. cos $\alpha = -0.9985$
25. sec $\alpha = 1.111$
26. csc $\alpha = 1.042$

Find all angles α such that $0 \leq \alpha < 2\pi$ and

27. cos $\alpha = 0.7452$
28. sin $\alpha = 0.3523$
29. tan $\alpha = -1.313$
30. cot $\alpha = -0.1001$

Using Table I in Appendix D, find all angles α such that

31. sin $\alpha = 0.7193$
32. tan $\alpha = 0.6330$
33. cot $\alpha = 1.600$
34. cos $\alpha = 0.9555$
35. cos $\alpha = -0.4990$
36. csc $\alpha = -1.234$

Using Table II in Appendix D, find all angles α such that

37. tan $\alpha = 2.360$
38. cos $\alpha = 0.7900$
39. sin $\alpha = -0.4350$
40. cot $\alpha = -1.700$

2.4 TRIGONOMETRIC FUNCTIONS AND THEIR GRAPHS

A real valued FUNCTION is a set of ordered pairs of real numbers (x, y) in which no two distinct ordered pairs have the same first element. The DOMAIN of the function is the set of first elements, x, and the RANGE of the function is the set of second elements, y. We have consistently referred to the relationships between sides of angles in previous sections as trigonometric ratios, due to our initial definitions, but it should be apparent that it is also possible to regard sine, cosine and the others as functions, if we make the following assumptions:

1. $\alpha \in R$, the set of real numbers, and α is the radian measure of an angle, or
2. $\alpha \in R$ and $m°(\alpha)$ is the degree measure of the angle.

For instance, the set of ordered pairs, $(\alpha, \sin \alpha)$, is a function because for each first element α there is a unique second element, $\sin \alpha$. The domain of the function is the set R of real numbers, and the range of the function is the set of numbers between -1 and 1, inclusive ($-1 \leq \sin \alpha \leq 1$). We can see this from the definitions because $x^2 + y^2 = r^2$, which means $y^2 \leq r^2$, and r is always positive, so $|y| \leq r$ and $|y|/r \leq 1$, or $-1 \leq y/r \leq 1$.

On the other hand the function $\{(\alpha, \tan \alpha)\}$ has as its domain the set $\{\alpha \mid \alpha \in R$ and $\alpha \neq (\pi/2 + k\pi)\}$. This is because $\tan \alpha = y/x$ by definition, and when $\alpha = \pi/2 + k\pi$ the denominator, x, is zero. Also when $\alpha = \pi/2 + k\pi$ the value of x is very close to zero while the value of y is close to r. A look at Table II in Appendix D for $\alpha = 1.55$ and 1.56 shows that $\tan \alpha$ grows very large. (In calculus we say that as α approaches $\pi/2$, $\tan \alpha$ approaches infinity.)

The cosecant function, $\{(\alpha, \csc \alpha)\}$, has as its domain the set $\{\alpha \mid \alpha \in R$ and $\alpha \neq k \cdot \pi\}$, and its range is $\{\csc \alpha \mid \csc \alpha \leq -1\}$ or $\{\csc \alpha \mid \csc \alpha \geq 1\}$. This is evident when we consider that $\csc \alpha = r/y$, and that $y = 0$ when $\alpha = 0, \pi, 2\pi,$ and so on, and also that $|y| \leq r$.

Since we are about to discuss the graphs of the trigonometric functions, there are two additional preliminaries which will prove helpful in the task. First we can prove that

$$\cos(-\alpha) = \cos \alpha$$

$$\sec(-\alpha) = \sec \alpha$$

$$\sin(-\alpha) = -\sin \alpha$$

$$\csc(-\alpha) = -\csc \alpha$$

$$\tan(-\alpha) = -\tan \alpha$$

$$\cot(-\alpha) = -\cot \alpha$$

The first two of these equations show that cosine and secant are EVEN functions and all the others are ODD functions. A function is an even function if $f(-x) = f(x)$ for all x in its domain, and it is an odd function if $f(-x) = -f(x)$ for all x in its domain. It can also be shown that the graph of an even function is symmetric with respect to the y axis while an odd function is symmetric with respect to the origin.

To prove $\cos(-\alpha) = \cos \alpha$, and so on, we can refer to Figure 2–21. In either case we see that angles POR and ROQ are equal, so if $PO = r$ and PQ is perpendicular to the x axis at R, then triangle POR is congruent to triangle QOR because they are right triangles with an acute angle and a leg of one equal to an acute angle and leg of the other. Thus $PR = QR$ and $QO = PO = r$, so that in Figure 2–21a we have $PR = y$, $QR = |-y|$, and $OR = x$. Thus

$$\cos \alpha = x/r \quad \text{and} \quad \cos(-\alpha) = x/r$$

2.4 TRIGONOMETRIC FUNCTIONS AND THEIR GRAPHS

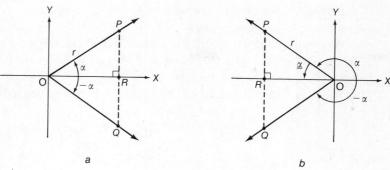

FIGURE 2-21

so $$\cos(-\alpha) = \cos\alpha$$

while $$\sin\alpha = y/r \quad \text{and} \quad \sin(-\alpha) = (-y)/r = -(y/r)$$

so $$\sin(-\alpha) = -\sin\alpha.$$

The same relationships hold in Figure 2–21b, and furthermore can be established even if α and $-\alpha$ are quadrantal angles so that there are no triangles involved.

The second preliminary referred to above is that the trigonometric functions are PERIODIC FUNCTIONS. This means that there is a number a such that $f(x + a) = f(x)$ for all x in the domain. We know that $\sin(x + 2\pi) = \sin x$, $\sin(x + 4\pi) = \sin x$, and so on, so $f(x) = \sin x$ is a periodic function. The smallest number a for which the above is true is called the PERIOD of the periodic function, and for the sine function it is 2π, while for the tangent function the period is π.

Now let us turn to the business of drawing the graphs. The domain of each of the trigonometric functions is infinite in extent, but because of their periodicity, we need draw them over only a portion of their domain. For $f(\alpha) = \sin\alpha$ we shall use the interval $-\pi/2 \leq \alpha \leq 5\pi/2$, which is approximately $-1.57 \leq \alpha \leq 7.85$. Recall that the range of $\sin\alpha$ is $-1 \leq \sin\alpha \leq 1$, and choose an appropriate scale. In Figure 2–22 we see the beginning of the graph, for which we have used Table II in Appendix D, and $\alpha = k(0.2)$ and also $\alpha = 1.57 \approx \pi/2$. For example, $\sin 0.8 = 0.7174$ according to the table, but rounding to $\sin 0.8 = 0.72$ is quite

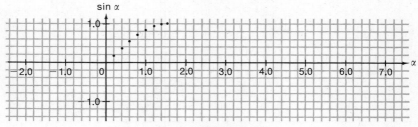

FIGURE 2-22

sufficient for our present purpose. Note that each square is equal to 0.2 units, so 5 squares = 1 unit, and so on.

At this stage the work begins in earnest. To find sin 1.6 we must use the reference angle and here $\alpha = 3.14 - 1.60 = 1.54$, so sin 1.6 = sin 1.54 = 0.9995. Next sin 1.80 = sin 1.34 = 0.9735, sin 2.00 = sin 1.14 = 0.9086, and so on, which we tabulate in Table 2–3. You will find that it will not be necessary to plot so many

TABLE 2–3

α	2.20	2.40	2.60	2.80	3.00	3.20
α	0.94	0.74	0.54	0.34	0.14	0.06
sin α	0.8076	0.6743	0.5141	0.3335	0.1395	−0.0600

points for other graphs. Observe that when $\alpha = 3.20$ it is in Quadrant III and the function value is negative. Plotting these additional points on the graph we have something which looks like Figure 2–23. Changing to an interval of 0.4 will be adequate for the remainder of the graph, and we also include points at $\alpha = 3\pi/2 \approx 4.71$, $\alpha = 2\pi \approx 6.28$, $\alpha = 5\pi/2 \approx 7.85$, and $\alpha = -\pi/2 \approx -1.57$ since they are maximum or minimum points or zeros of the function. The points in the interval $-1.57 \leq \alpha \leq 0$ are derived from the relation $\sin(-\alpha) = -\sin\alpha$, which means, for example, that if we want to find sin (−0.4), we simply find −sin 0.4 = −0.3894. The remaining points we have plotted are tabulated in Table 2–4, rounded off to two decimal places. The graph with all points plotted is shown in Figure 2–24.

In conclusion, it appears that values of α between those plotted should produce values of sin α such that the new points lie between the previously plotted ones. This means that we should be able to draw a smooth curve through these points to produce the graph shown in Figure 2–25.

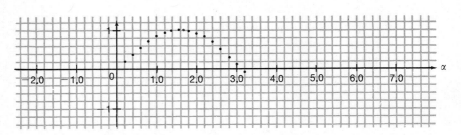

FIGURE 2–23

TABLE 2–4

α	3.40	3.80	4.20	4.60	5.00	5.40	5.80	6.20	6.60	7.00	7.40	7.80
α	0.26	0.66	1.06	1.46	1.28	0.88	0.48	0.08	0.32	0.72	1.12	1.52
sin α	−0.26	−0.61	−0.87	−0.99	−0.96	−0.77	−0.46	−0.08	0.31	0.66	0.90	1.00

2.4 TRIGONOMETRIC FUNCTIONS AND THEIR GRAPHS

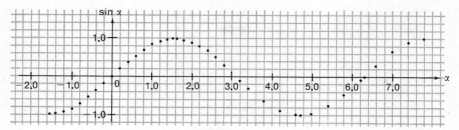

FIGURE 2–24

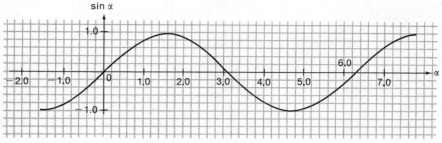

FIGURE 2–25

Plotting the graph carefully over the interval we used shows how the sine function repeats itself after going over its period, 2π. It would now be possible to draw the graph by plotting considerably fewer points.

Using essentially the same techniques, we illustrate next the graphs of $f(\alpha) = \tan \alpha$ and $f(\alpha) = \csc \alpha$ which may be seen in Figure 2–26 and Figure 2–27. The points are tabulated in Table 2–5. The task of plotting the graphs of the remaining trigonometric functions is left to you in the exercises.

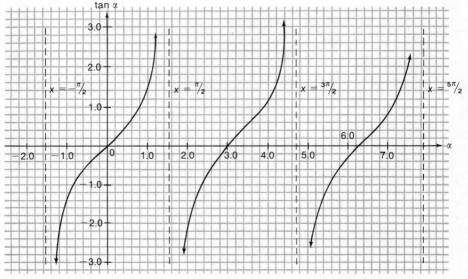

FIGURE 2–26

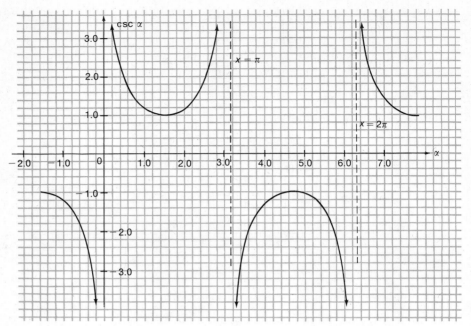

FIGURE 2–27

TABLE 2–5

α	−1.20	−0.80	−0.40	−0.00	0.40	0.80	1.20	1.60
α̲	1.20	0.80	0.40	0.00	0.40	0.80	1.20	1.54
tan α	−2.57	−1.03	−0.42	0.00	0.42	1.03	2.57	−32.46
csc α	−1.07	−1.39	−2.57	—	2.57	1.39	1.07	1.00
α	2.00	2.40	2.80	3.20	3.60	4.00	4.40	4.80
α̲	1.14	0.74	0.34	0.06	0.46	0.86	1.26	1.48
tan α	−2.18	−0.91	−0.35	0.06	0.50	1.16	3.11	−10.98
csc α	1.10	1.48	3.00	16.68	−2.09	−1.32	−1.05	−1.00
α	5.20	5.60	6.00	6.40	6.80	7.20	7.60	
α̲	1.08	0.68	0.28	0.12	0.52	0.92	1.32	
tan α	−1.87	−0.81	−0.29	0.12	0.57	1.29	3.90	
csc α	−1.13	−1.59	−3.62	8.35	1.85	1.26	1.03	

REVIEW EXERCISES

EXERCISES

Draw the graph of each of the following over $-\pi/2 \le \alpha \le 5\pi/2$. Show the table of values for each.

1. $f(\alpha) = \cos \alpha$
2. $f(\alpha) = \sec \alpha$
3. $f(\alpha) = \cot \alpha$
4. Draw a figure with $\alpha = \pi/2$ and $-\alpha = -\pi/2$ and show that $\cos(-\alpha) = \cos \alpha$, $\sin(-\alpha) = -\sin \alpha$, and so on.
5. Draw the graphs of each of the following functions and tell whether they are odd, even, or neither.
   a. $f(x) = x^2$
   b. $f(x) = x^3$
   c. $f(x) = x^2 + x$
   d. $f(x) = 4 - x^2$
6. Demonstrate why an even function is symmetric with respect to the y axis. *Hint:* Refer to the definition of an even function. Consider a function f for which $f(3) = 5$, $f(-3) = 5$, $f(2) = 1$, $f(-2) = 1$, $f(1) = 0$, $f(-1) = 0$. Plot these points, then generalize. What is the meaning of symmetry with respect to a line?

REVIEW EXERCISES

1. Find the approximate value of each of the following, using Table I in Appendix D:
   a. $\cos 53°40'$
   b. $\cot 310°$
   c. $\sin 160°$
   d. $\tan 820°20'$
   e. $\sec 245°30'$
   f. $\sin(-35°50')$
   g. $\tan(-215°)$
   h. $\cot(-100°40')$

2. Find the approximate value of each of the following using Table II in Appendix D, and $\pi \approx 3.14$ or $2\pi \approx 6.28$ where appropriate:
   a. $\sin 1.00^R$
   b. $\tan 2.60^R$
   c. $\cos 5.00^R$
   d. $\cot 7.50^R$

3. Find all angles α such that $0° \le \alpha < 360°$ and
   a. $\cot \alpha = 0.7133$
   b. $\sin \alpha = -0.3800$
   c. $\cos \alpha = -0.5100$
   d. $\tan \alpha = 1.303$

4. Find all angles α such that $0^R \le \alpha < 2\pi^R$ and
   a. $\cos \alpha = 0.7900$
   b. $\tan \alpha = -2.066$

5. Given that the terminal side of α passes through the point $(5, -12)$, find all six trigonometric ratios for α.

6. If $\cos \alpha = -2\sqrt{2}/3$ and α terminates in Quadrant II, find the other five trigonometric ratios for α.

7. Find the exact value of each of the following (do not use tables):
   a. $\tan 300°$
   b. $\sin(5\pi/4)^R$
   c. $\cos 390°$
   d. $\cot(-\pi)^R$

8. Sketch the graph of each of the following over the interval $0^R \le \alpha < 2\pi^R$ indicating the location of zeros, maximum and minimum points, and asymptotes if any.
   a. $f(\alpha) = \cos \alpha$
   b. $f(\alpha) = \tan \alpha$

CHAPTER THREE

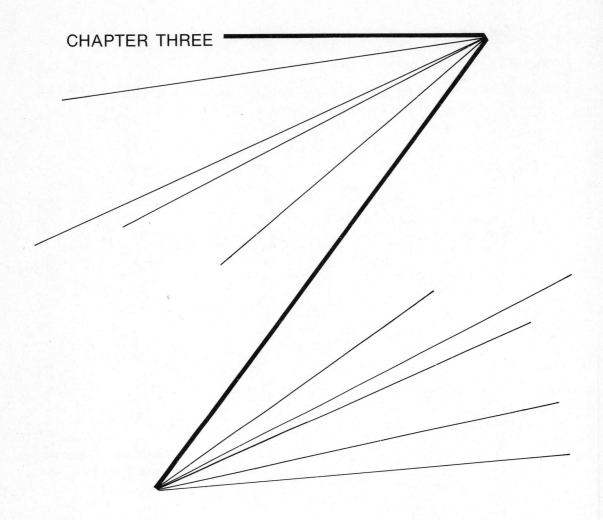

Identities, Formulas, and Equations

3.1 PROVING IDENTITIES

We have already proved a number of trigonometric identities by means of the definitions of the trigonometric ratios. Now we are ready to expand our capabilities in this area, obtaining skills which are necessary in the development of certain formulas in science and for many simplification techniques in advanced mathematics.

3.1 PROVING IDENTITIES

The identities we have already established are the RECIPROCAL IDENTITIES

$$\sin \alpha \csc \alpha = 1 \tag{1}$$

$$\cos \alpha \sec \alpha = 1 \tag{2}$$

$$\tan \alpha \cot \alpha = 1, \tag{3}$$

and the QUOTIENT IDENTITIES

$$\tan \alpha = \frac{\sin \alpha}{\cos \alpha} \tag{4}$$

$$\cot \alpha = \frac{\cos \alpha}{\sin \alpha} \tag{5}$$

As a reminder of the method of proving the identities by means of the definitions, we recall that $\cot \alpha = x/y$, $\sin \alpha = y/r$, and $\cos \alpha = x/r$, so

$$\frac{\cos \alpha}{\sin \alpha} = \frac{x/r}{y/r} = \frac{x}{y} = \cot \alpha$$

It should be noted that when we speak of a trigonometric identity we mean that the two members of the equation are equal for every value of the variable for which both members are defined. The last identity is not defined, for example, when $\sin \alpha = 0$, or when $\alpha = k\pi$.

Another set of identities provable by means of the definitions are the PYTHAGOREAN IDENTITIES, the first of which is

$$\sin^2 \alpha + \cos^2 \alpha = 1* \tag{6}$$

We can prove this by writing

$$\sin^2 \alpha + \cos^2 \alpha = \left(\frac{y}{r}\right)^2 + \left(\frac{x}{r}\right)^2 = \frac{y^2}{r^2} + \frac{x^2}{r^2}$$

$$= \frac{y^2 + x^2}{r^2} = \frac{r^2}{r^2} = 1$$

since $y^2 + x^2 = x^2 + y^2 = r^2$. The other two Pythagorean identities are

$$1 + \tan^2 \alpha = \sec^2 \alpha \tag{7}$$

$$\cot^2 \alpha + 1 = \csc^2 \alpha \tag{8}$$

* $\sin^2 \alpha = (\sin \alpha)^2$, as explained in Section 2.2.

Both could be proved by means of the definitions, but we need to develop another method of proof. Let us agree that any natural number power of an identity is also an identity. In other words, since $\tan \alpha = \sin \alpha / \cos \alpha$, it follows that $\tan^2 \alpha = (\sin \alpha / \cos \alpha)^2 = \sin^2 \alpha / \cos^2 \alpha$, $\tan^3 \alpha = \sin^3 \alpha / \cos^3 \alpha$, and so on. Also let us agree that since $\sin \alpha \csc \alpha = 1$, it follows that $\csc \alpha = 1/\sin \alpha$ ($\sin \alpha \neq 0$), or $\sin \alpha = 1/\csc \alpha$, and similarly for the other reciprocal identities. Since $\sin^2 \alpha + \cos^2 \alpha = 1$, we can divide both members of the equation by $\cos^2 \alpha$ to obtain

$$\frac{\sin^2 \alpha + \cos^2 \alpha}{\cos^2 \alpha} = \frac{1}{\cos^2 \alpha}$$

or

$$\frac{\sin^2 \alpha}{\cos^2 \alpha} + \frac{\cos^2 \alpha}{\cos^2 \alpha} = \frac{1}{\cos^2 \alpha}$$

from which

$$\tan^2 \alpha + 1 = \sec^2 \alpha$$

Dividing both members of identity (6) by $\sin^2 \alpha$ will produce identity (8).

The eight identities established above are referred to as the FUNDAMENTAL IDENTITIES. With these eight we can prove a large number of other identities, which will be the main business of this chapter. Although the proving of identities is a useful and worthwhile skill in itself, it can also be treated as a game which tends to make it (and many other kinds of work) more palatable. The rules of the game require that the players change one member of the identity into the other by means of algebraic substitutions of known identities. There are other ways of playing the game which will be mentioned later, but this form produces many beneficial results and avoids certain pitfalls.

EXAMPLE 1   Prove that $\csc \alpha = \dfrac{\cot \alpha}{\cos \alpha}$ is an identity.

SOLUTION   There are usually many ways to prove an identity. The simplest way is generally the best way, although it is sometimes hard to decide which method is simplest. If we start with the right member, however,

$$\frac{\cot \alpha}{\cos \alpha} = \cot \alpha \cdot \frac{1}{\cos \alpha} = \frac{\cos \alpha}{\sin \alpha} \cdot \frac{1}{\cos \alpha} = \frac{1}{\sin \alpha} = \csc \alpha$$

We consider as "obvious" the replacements $\dfrac{\cos \alpha}{\sin \alpha}$ for $\cot \alpha$ and $\csc \alpha$ for $\dfrac{1}{\sin \alpha}$. Sometimes other slight variations on the fundamental identities are used in addition to those which have already been mentioned. For example, from (6) we have $\sin^2 \alpha = 1 - \cos^2 \alpha$ and $\cos^2 \alpha = 1 - \sin^2 \alpha$, from (7) we have $\tan^2 \alpha = \sec^2 \alpha - 1$ and $\sec^2 \alpha - \tan^2 \alpha = 1$, and from (8) we know that $\cot^2 \alpha = \csc^2 \alpha - 1$ and $\csc^2 \alpha - \cot^2 \alpha = 1$.

3.1 PROVING IDENTITIES

EXAMPLE 2 Prove that $\dfrac{1 - \sin^2 \alpha}{\sin^2 \alpha} = \cot^2 \alpha$ is an identity.

SOLUTION Here we could proceed in two related ways. The first is to replace $1 - \sin^2 \alpha$ with $\cos^2 \alpha$, so that

$$\frac{1 - \sin^2 \alpha}{\sin^2 \alpha} = \frac{\cos^2 \alpha}{\sin^2 \alpha} = \cot^2 \alpha$$

using a power of identity (5). The second is to divide through first by $\sin^2 \alpha$, so that

$$\frac{1 - \sin^2 \alpha}{\sin^2 \alpha} = \frac{1}{\sin^2 \alpha} - \frac{\sin^2 \alpha}{\sin^2 \alpha} = \csc^2 \alpha - 1 = \cot^2 \alpha$$

The second solution is probably not as simple as the first, but it is equally valid and may seem to some to be the obvious thing to do.

Although there is no set procedure for proving identities, it is sometimes useful (especially when confronted with one for which there seems to be no better way) to change everything into terms of sines and cosines and go on from there.

EXAMPLE 3 Prove that $\tan \alpha + \cot \alpha = \sec \alpha \csc \alpha$ is an identity.

SOLUTION Change $\tan \alpha$ and $\cot \alpha$ to equivalent expressions in $\sin \alpha$ and $\cos \alpha$, then note that the right member is a single term, which motivates us to combine the two fractions.

$$\tan \alpha + \cot \alpha = \frac{\sin \alpha}{\cos \alpha} + \frac{\cos \alpha}{\sin \alpha}$$

$$= \frac{\sin \alpha}{\cos \alpha} \cdot \frac{\sin \alpha}{\sin \alpha} + \frac{\cos \alpha}{\sin \alpha} \cdot \frac{\cos \alpha}{\cos \alpha}$$

$$= \frac{\sin^2 \alpha + \cos^2 \alpha}{\cos \alpha \cdot \sin \alpha} = \frac{1}{\cos \alpha \cdot \sin \alpha}$$

$$= \frac{1}{\cos \alpha} \cdot \frac{1}{\sin \alpha} = \sec \alpha \csc \alpha.$$

The replacement of $\sin^2 \alpha + \cos^2 \alpha$ by 1 which took place halfway through will become obvious in time. Perhaps the most important thing to do is to look ahead to see where we are going, in this case causing us to add the two fractions. (Of course we have to remember how to add fractions. Don't forget that sin α, cos α, and so on are just numbers or variables like *m* and *n*, *x* and *y*.)

Since proving identities often takes a great deal of writing and space, let us agree to use LM to stand for the left member or RM in place of the right member when we begin our proof.

EXAMPLE 4   Prove that $\dfrac{1 - \tan^2 \alpha}{\tan \alpha} = \cot \alpha - \tan \alpha$ is an identity.

SOLUTION $LM = \dfrac{1}{\tan \alpha} - \tan \alpha = \cot \alpha - \tan \alpha$

We could use

$$RM = \dfrac{1}{\tan \alpha} - \dfrac{\tan \alpha}{1} \cdot \dfrac{\tan \alpha}{\tan \alpha} = \dfrac{1 - \tan^2 \alpha}{\tan \alpha}$$

but this seems a trifle more difficult.

It was observed earlier that we could return to the definitions to prove the identities. The last example, for instance, would be true since

$$\dfrac{1 - \tan^2 \alpha}{\tan \alpha} = \dfrac{1 - \left(\dfrac{y}{x}\right)^2}{\dfrac{y}{x}} = \dfrac{1 - \dfrac{y^2}{x^2}}{\dfrac{y}{x}} \cdot \dfrac{\dfrac{x}{y}}{\dfrac{x}{y}}$$

$$= \dfrac{\dfrac{x}{y} - \dfrac{y}{x}}{1} = \dfrac{x}{y} - \dfrac{y}{x} = \cot \alpha - \tan \alpha$$

But this is cumbersome and certainly will not be as useful later on as full use of the identities will be.

Yet another proof of the last example could be argued by multiplying both sides of the "equation" by tan α. In other words, if

$$\dfrac{1 - \tan^2 \alpha}{\tan \alpha} = \cot \alpha - \tan \alpha$$

$$\dfrac{\tan \alpha}{1} \cdot \dfrac{1 - \tan^2 \alpha}{\tan \alpha} = \tan \alpha \, (\cot \alpha - \tan \alpha)$$

so $1 - \tan^2 \alpha = \tan \alpha \cdot \cot \alpha - \tan^2 \alpha$

3.1 PROVING IDENTITIES

or
$$1 - \tan^2 \alpha = 1 - \tan^2 \alpha$$

The fallacy in the above is that it assumes the truth of that which is to be proved. If it can be shown that the steps are reversible, on the other hand, the proof is valid. Again it seems a bit more sporting to change one member into the other, and this is what we shall do in all our work in this chapter. It *is* difficult to decide how to proceed sometimes, as the next example will demonstrate.

EXAMPLE 5 Prove that $\dfrac{1 - \cos \alpha}{\sin \alpha} = \dfrac{1}{\csc \alpha + \cot \alpha}$

SOLUTION Here it appears that one member is as complicated as the other. For instance,

$$\text{LM} = \frac{1}{\sin \alpha} - \frac{\cos \alpha}{\sin \alpha} = \csc \alpha - \cot \alpha$$

$$\text{RM} = \frac{1}{\dfrac{1}{\sin \alpha} + \dfrac{\cos \alpha}{\sin \alpha}} \cdot \frac{\sin \alpha}{\sin \alpha} = \frac{\sin \alpha}{1 + \cos \alpha}$$

Either way seems to run into difficulty. If we persevere, however, it can be done. Observe that the LM approach has indeed led us into an expression in csc α and cot α. Since the right member has csc α + cot α as its denominator, let us *force* csc α + cot α into the denominator of csc α − cot α

$$\text{LM} = \frac{1}{\sin \alpha} - \frac{\cos \alpha}{\sin \alpha} = \csc \alpha - \cot \alpha = \frac{\csc \alpha - \cot \alpha}{1} \cdot \frac{\csc \alpha + \cot \alpha}{\csc \alpha + \cot \alpha}$$

$$= \frac{\csc^2 \alpha - \cot^2 \alpha}{\csc \alpha + \cot \alpha} = \frac{1}{\csc \alpha + \cot \alpha}.$$

It may seem unlikely that we would remember that $\csc^2 \alpha - \cot^2 \alpha = 1$, but that's what the game is all about, and besides we can peek at RM without it being considered cheating. If we choose to operate on RM all the way, we can still complete the proof, although with a bit more difficulty. Starting where we left off above,

$$\text{RM} = \frac{\sin \alpha}{1 + \cos \alpha} \cdot \frac{1 - \cos \alpha}{1 - \cos \alpha} = \frac{\sin \alpha (1 - \cos \alpha)}{1 - \cos^2 \alpha}$$

$$= \frac{\sin \alpha (1 - \cos \alpha)}{\sin^2 \alpha} = \frac{\sin \alpha (1 - \cos \alpha)}{\sin \alpha \cdot \sin \alpha}$$

$$= \frac{1 - \cos \alpha}{\sin \alpha}$$

So far we have been letting α stand for the measure of the angle, but any letter will do. One final example before you are invited to try the game yourself.

EXAMPLE 6 Prove that $\sin t\,(1 + \tan^2 t) - \sin t = \cos t \tan^3 t$ is an identity.

SOLUTION $\text{LM} = \sin t + \sin t \tan^2 t - \sin t = \sin t \tan^2 t$

Looking at RM we see we need the *cube* of tan t, so we can get it if we multiply by $\cot t \tan t = 1$:

$$\sin t \tan^2 t = \sin t\,(\cot t \tan t)\tan^2 t$$

$$= \sin t \left(\frac{\cos t}{\sin t}\right) \tan^3 t = \cos t \tan^3 t$$

EXERCISES

Prove that each of the following is an identity.

1. $\sin \alpha \sec \alpha = \tan \alpha$
2. $\csc \alpha \cos \alpha = \cot \alpha$
3. $\sin^2 \alpha\,(\csc^2 \alpha - 1) = \cos^2 \alpha$
4. $\tan^2 \alpha\,(1 + \cot^2 \alpha) = \sec^2 \alpha$
5. $\dfrac{\tan u}{\sec u} = \sin u$
6. $\dfrac{\sec v}{\tan v} = \csc v$
7. $\sin^2 \alpha\,(1 + \cot^2 \alpha) = 1$
8. $\sec^2 \alpha\,(1 - \sin^2 \alpha) = 1$
9. $\dfrac{\sec t}{\cos t} = 1 + \dfrac{\tan t}{\cot t}$
10. $\dfrac{\csc u}{\sin u} - \dfrac{\cot u}{\tan u} = 1$
11. $\dfrac{1}{\tan \theta + \cot \theta} = \sin \theta \cos \theta$
12. $\dfrac{1}{\sec x - \cos x} = \cot x \csc x$
13. $\dfrac{\sin u \sec u}{\tan u + \cot u} = \sin^2 u$
14. $\dfrac{\tan G + \sin G}{\csc G + \cot G} = \sin^2 G \sec G$
15. $(\tan y + \cot y)^2 = \sec^2 y + \csc^2 y$
16. $(\tan y + \cot y)^2 = \sec^2 y \csc^2 y$
17. $\dfrac{\sec^2 \alpha}{\sec^2 \alpha - 1} = \csc^2 \alpha$
18. $\dfrac{1 + \sin^2 \alpha \sec^2 \alpha}{1 + \cos^2 \alpha \csc^2 \alpha} = \tan^2 \alpha$
19. $\dfrac{\sin A + \cos A}{\sec A + \csc A} = \dfrac{\sin A}{\sec A}$
20. $\dfrac{\sec B + \csc B}{1 + \tan B} = \dfrac{\cot B}{\cos B}$
21. $\sec^4 \alpha - \sec^2 \alpha = \tan^4 \alpha + \tan^2 \alpha$
22. $\cos^4 x - \sin^4 x = 1 - 2\sin^2 x$
23. $\dfrac{1}{1 + \sin A} + \dfrac{1}{1 - \sin A} = 2 \sec^2 A$
24. $\dfrac{1}{1 - \cos B} + \dfrac{1}{1 + \cos B} = 2 \csc^2 B$
25. $\dfrac{1 - \cos \theta}{1 + \cos \theta} = (\cot \theta - \csc \theta)^2$
26. $\dfrac{1 + \sin u}{1 - \sin u} = (\sec u + \tan u)^2$

27. $\dfrac{\sin \alpha - \cos \alpha}{\sin \alpha + \cos \alpha} = \dfrac{\tan \alpha - 1}{\tan \alpha + 1}$

28. $\dfrac{\sec \alpha - \csc \alpha}{\sec \alpha + \csc \alpha} = \dfrac{\tan \alpha - 1}{\tan \alpha + 1}$

29. $\cos^2 x \cot^2 x + \sin^2 x \tan^2 x + 1 = \tan^2 x + \cot^2 x$

30. $\sin y \tan^2 y + \csc y \sec^2 y - 2 \tan y \sec y = \csc y - \sin y$

31. $\dfrac{1 + \cos u}{\sin u} = \dfrac{\sin u}{1 - \cos u}$

32. $\dfrac{1 - \sin u}{\cos u} = \dfrac{\cos u}{1 + \sin u}$

33. $\dfrac{1 + \sin \alpha}{1 - \sin \alpha} - \dfrac{1 - \sin \alpha}{1 + \sin \alpha} = 4 \sec \alpha \tan \alpha$

34. $\dfrac{1 - \cos \alpha}{1 + \sin \alpha} + \dfrac{1 + \cos \alpha}{1 - \sin \alpha} = 2(\tan^2 \alpha + \tan \alpha + 1)$

35. $\dfrac{\tan x - \cot x}{\sec x} = \dfrac{1 - 2 \cos^2 x}{\sin x}$

36. $\dfrac{\tan u - \sin u}{\sec u} = \dfrac{\sin^3 u}{1 + \cos u}$

3.2 ADDITION AND SUBTRACTION IDENTITIES

From time to time you will need to find the trigonometric ratios for the sum of two angles. It might seem that $\cos (60° + 30°)$ should equal $\cos 60° + \cos 30°$, but it doesn't. To show that $\cos (60° + 30°) \neq \cos 60° + \cos 30°$, we simply observe that $\cos (60° + 30°) = \cos 90° = 0$, while $\cos 60° + \cos 30° = \frac{1}{2} + \frac{1}{2}\sqrt{3}$, which is certainly not 0. In order to write $\cos 90°$ in terms of $60°$ and $30°$ ratios we must use $\cos (60° + 30°) = \cos 60° \cos 30° - \sin 60° \sin 30°$, a somewhat more complicated expression but correct since $\cos 60° \cos 30° - \sin 60° \sin 30° = \frac{1}{2} \cdot \frac{1}{2}\sqrt{3} - \frac{1}{2}\sqrt{3} \cdot \frac{1}{2} = 0$. A formula for determining $\cos (\alpha + \beta)$ is

$$\cos (\alpha + \beta) = \cos \alpha \cos \beta - \sin \alpha \sin \beta$$

which we shall prove valid for all angles α and β a bit later on.

The formula just stated is one of six addition and subtraction formulas to be developed. These in turn will lead to some other special formulas or identities, so that there will be a total of 20 formulas, including the eight fundamental identities, which you will have at your command by the end of this chapter. In order to prove the formula for $\cos (\alpha + \beta)$ given above, it will be necessary to use a combination of algebra, geometry and some previously proven formulas of immense importance for the development of trigonometry and its later applications. First let us recall that we have already established (in Section 2.4) that

$$\sin (-\alpha) = -\sin \alpha$$

$$\cos (-\alpha) = \cos \alpha$$

$$\tan (-\alpha) = -\tan \alpha$$

Now consider two angles α and β which are situated as in Figure 3–1a, with $\alpha > \beta$. The circle has radius 1 unit, and $m(\angle POQ) = m(\alpha - \beta)$. If we construct angle $P'O'Q'$ as in Figure 3–1b, so that $m(\angle P'O'Q') = m(\alpha - \beta)$, then if the radius of circle O' is also 1, $P'Q' = PQ$. By the distance formula we have

$$PQ = \sqrt{(x_2 - x_1)^2 + (y_2 - y_1)^2}$$
$$P'Q' = \sqrt{(x_3 - 1)^2 + (y_3 - 0)^2}.$$

Squaring both sides of the last two equations, we have

$$PQ^2 = x_2^2 - 2x_2 x_1 + x_1^2 + y_2^2 - 2y_2 y_1 + y_1^2$$

and $$P'Q'^2 = x_3^2 - 2x_3 + 1 + y_3^2$$

Since $x_1^2 + y_1^2 = x_2^2 + y_2^2 = x_3^2 + y_3^2 = 1$, we can simplify the above to

$$PQ^2 = 2 - 2x_2 x_1 - 2y_2 y_1$$

and $$P'Q'^2 = 2 - 2x_3$$

Then $$P'Q'^2 = PQ^2$$

so $$2 - 2x_3 = 2 - 2x_2 x_1 - 2y_2 y_1$$

or $$-2x_3 = -2x_2 x_1 - 2y_2 y_1$$

and $$x_3 = x_2 x_1 + y_2 y_1$$

This has been quite a bit of algebra, but now if we refer again to Figure 3–1 we see that $\cos \beta = x_1/1 = x_1$, $\sin \beta = y_1/1 = y_1$, and similarly $x_2 = \cos \alpha$, $y_2 = \sin \alpha$, and $x_3 = \cos(\alpha - \beta)$. Substituting into the last equation of our sequence above, we obtain

$$\cos(\alpha - \beta) = \cos \alpha \cos \beta + \sin \alpha \sin \beta \tag{1}$$

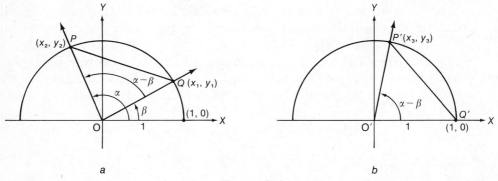

Figure 3–1

3.2 ADDITION AND SUBTRACTION IDENTITIES

This is not the formula we set out to establish, but it *is* valid for all α and β, so

$$\cos(\alpha + \beta) = \cos[\alpha - (-\beta)] = \cos\alpha\cos(-\beta) + \sin\alpha\sin(-\beta)$$
$$= \cos\alpha\cos\beta + \sin\alpha(-\sin\beta)$$
$$\cos(\alpha + \beta) = \cos\alpha\cos\beta - \sin\alpha\sin\beta \tag{2}$$

We might point out before leaving this proof that we could have substituted $\cos\alpha$ and $\sin\alpha$ for x_1 and y_1, and so on, immediately, and used $\sin^2\alpha + \cos^2\alpha = 1$ in the proof. The reader may wish to try it this way.

To develop similar formulas for $\sin(\alpha + \beta)$ and $\sin(\alpha - \beta)$ is not so difficult, although some additional preliminaries are required. First we establish that

$$\cos\left(\frac{\pi}{2} - \alpha\right) = \sin\alpha \tag{3}$$

from equation (2), above, since

$$\cos\left(\frac{\pi}{2} - \alpha\right) = \cos\frac{\pi}{2}\cos\alpha + \sin\frac{\pi}{2}\sin\alpha$$
$$= 0 \cdot \cos\alpha + 1 \cdot \sin\alpha = \sin\alpha$$

Equation (3) is true for all values of α, so $\sin\beta = \cos\left(\frac{\pi}{2} - \beta\right)$, and if β is replaced by $\frac{\pi}{2} - \alpha$, we obtain

$$\sin\left(\frac{\pi}{2} - \alpha\right) = \cos\left[\frac{\pi}{2} - \left(\frac{\pi}{2} - \alpha\right)\right]$$
$$\sin\left(\frac{\pi}{2} - \alpha\right) = \cos\alpha \tag{4}$$

You will recall that we proved the special cases of equations (3) and (4) for α an acute angle back in Section 1.2.

Now we use equation (3) to write

$$\sin(\alpha - \beta) = \cos\left[\frac{\pi}{2} - (\alpha - \beta)\right]$$
$$= \cos\left[\left(\frac{\pi}{2} - \alpha\right) + \beta\right]$$

then substituting from equation (2) we write

$$\sin(\alpha - \beta) = \cos\left(\frac{\pi}{2} - \alpha\right)\cos\beta - \sin\left(\frac{\pi}{2} - \alpha\right)\sin\beta$$

and substituting from equations (3) and (4)

$$\sin(\alpha - \beta) = \sin\alpha\cos\beta - \cos\alpha\sin\beta \tag{5}$$

To obtain the corresponding formula for $\sin(\alpha + \beta)$ we write $\sin[\alpha - (-\beta)]$ and derive

$$\sin(\alpha + \beta) = \sin\alpha\cos\beta + \cos\alpha\sin\beta \tag{6}$$

the details of which we leave as an exercise.

If we know $\sin(\alpha + \beta)$ and $\cos(\alpha + \beta)$, it is easy to find $\tan(\alpha + \beta)$, since $\tan(\alpha + \beta) = \dfrac{\sin(\alpha + \beta)}{\cos(\alpha + \beta)}$. If we have $\tan\alpha$ and $\tan\beta$, however, we may want to use the formula

$$\tan(\alpha + \beta) = \frac{\tan\alpha + \tan\beta}{1 - \tan\alpha\tan\beta} \tag{7}$$

You could prove that equation (7) is an identity by the same methods used in the previous section, that is,

$$\text{RM} = \frac{\dfrac{\sin\alpha}{\cos\alpha} + \dfrac{\sin\beta}{\cos\beta}}{1 - \dfrac{\sin\alpha}{\cos\alpha}\cdot\dfrac{\sin\beta}{\cos\beta}} \cdot \frac{\cos\alpha\cos\beta}{\cos\alpha\cos\beta}$$

$$= \frac{\sin\alpha\cos\beta + \cos\alpha\sin\beta}{\cos\alpha\cos\beta - \sin\alpha\sin\beta}$$

$$= \frac{\sin(\alpha + \beta)}{\cos(\alpha + \beta)} = \tan(\alpha + \beta)$$

Reversing the steps in the above proof would constitute a derivation of equation (7), but it would be difficult to decide what to derive if the reader started out with $\tan(\alpha + \beta)$. The subtraction formula for tangent is

$$\tan(\alpha - \beta) = \frac{\tan\alpha - \tan\beta}{1 + \tan\alpha\tan\beta} \tag{8}$$

Since it is desirable to memorize the equations which have been established thus far and they are a bit confusing with the changes in algebraic signs, it may help to write

$$\sin(\alpha \pm \beta) = \sin\alpha\cos\beta \pm \cos\alpha\sin\beta$$

$$\cos(\alpha \pm \beta) = \cos\alpha\cos\beta \mp \sin\alpha\sin\beta$$

3.2 ADDITION AND SUBTRACTION IDENTITIES

$$\tan(\alpha \pm \beta) = \frac{\tan \alpha \pm \tan \beta}{1 \mp \tan \alpha \tan \beta}$$

if we use upper signs on both sides or lower signs on both sides. (The $\mp$ is called a minus or plus sign.)

You may wonder where all this is leading—why bother with all of these complicated formulas? It might be interesting to consider a few applications. In Appendix B we present a method for computing the values of the trigonometric functions $\sin \alpha$ and $\cos \alpha$ when α is in radians. Before the days of the high speed electronic computer it was an extremely slow and tedious process. If you refer to the formulas in Appendix B, you would probably agree that it would be easier to obtain $\sin 0.43$ by using $\sin 0.43 = \sin(0.4 + 0.03) = \sin 0.4 \cos 0.03 + \cos 0.4 \sin 0.03$ rather than

$$\sin 0.43 = 0.43 - \frac{(0.43)^3}{3!} + \frac{(0.43)^5}{5!} - \frac{(0.43)^7}{7!} + \frac{(0.43)^9}{9!} - \cdots$$

assuming that the values for 0.4 and 0.03 had already been worked out. There are also certain combinations which will produce *exact* values.

EXAMPLE 1 Find the exact value of $\cos \frac{5\pi}{12}$.

SOLUTION Since $5\pi/12$ can be written as the sum $(3\pi/12) + (2\pi/12)$

$$\cos \frac{5\pi}{12} = \cos\left(\frac{\pi}{4} + \frac{\pi}{6}\right)$$

$$= \cos \frac{\pi}{4} \cos \frac{\pi}{6} - \sin \frac{\pi}{4} \sin \frac{\pi}{6}$$

$$= \tfrac{1}{2}\sqrt{2} \cdot \tfrac{1}{2}\sqrt{3} - \tfrac{1}{2}\sqrt{2} \cdot \tfrac{1}{2}$$

$$= \tfrac{1}{4}(\sqrt{6} - \sqrt{2})$$

The trick is to find a combination of the special angles of Section 2.2 which have a sum or difference equal to the angle to be found.

There are times when we might need to know the angle between two lines whose equations are known. Suppose we have two lines ℓ_1 and ℓ_2, whose inclinations (the least positive angle the line makes with the x axis) are α_1 and α_2, respectively. If $\alpha_1 < \alpha_2$ and ϕ (Greek letter phi) is the angle between the lines such that $\phi = \alpha_2 - \alpha_1$ (refer to Figure 3–2),

$$\tan \phi = \tan(\alpha_2 - \alpha_1) = \frac{\tan \alpha_2 - \tan \alpha_1}{1 + \tan \alpha_2 \tan \alpha_1}$$

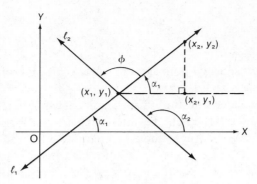

FIGURE 3–2

If (x_1, y_1) and (x_2, y_2) are points on a line, its slope, m, is found by the formula

$$m = \frac{y_2 - y_1}{x_2 - x_1}$$

Thus $m_1 = \tan \alpha_1$ and $m_2 = \tan \alpha_2$, so $\tan \phi = \dfrac{m_2 - m_1}{1 + m_1 m_2}$.

EXAMPLE 2 Find, to the nearest degree, the greater angle between two lines whose equations are $3x + 2y - 14 = 0$ and $3x - y - 11 = 0$.

SOLUTION We need the slopes of the two lines. Changing the equations to slope-intercept form we obtain $y = -\dfrac{3}{2}x + 7$ for the first and $y = 3x - 11$ for the second, so the slopes are $m_1 = 3$ and $m_2 = -\dfrac{3}{2}$. We assumed that $\alpha_1 < \alpha_2$, so $m_1 = 3$ implies that $0 < \alpha_1 < 90°$, while $m_2 = -\dfrac{3}{2}$ implies that $90° < \alpha_2 < 180°$. Thus

$$\tan \phi = \frac{-\dfrac{3}{2} - 3}{1 + 3\left(-\dfrac{3}{2}\right)} = \frac{-3 - 6}{2 - 9}$$

$$= \frac{-9}{-7} \approx 1.286$$

and $\phi \approx 52°10'$. The *greater* angle between the two lines, however, would be the supplement of ϕ, $180° - 52°10' = 127°50'$.

The following is not exactly an application of one of the addition formulas, but shows that skills derived from manipulation of these formulas and equations can be useful in an area which might be unexpected. In a book on machinery design, M. F. Spotts presents a formula for calculating torque in power screws:

$$T = r_t W \left[\frac{\cos \theta_n \sin \alpha + \mu_1 \cos \alpha}{\cos \theta_n \cos \alpha - \mu_1 \sin \alpha} + \frac{r_c}{r_t} \mu_2 \right]$$

which can be simplified to

$$T = r_t W \left[\frac{\cos \theta_n \tan \alpha + \mu_1}{\cos \theta_n - \mu_1 \tan \alpha} + \frac{r_c}{r_t} \mu_2 \right]^*$$

(We don't attempt to define the variables above, but rather leave that task to Mr. Spotts.) Part of the derivation of the addition formula for tangents is very much like the simplified equation above, so learning the derivations in this section may prove to be useful.

EXERCISES

1. Derive the equation for $\sin(\alpha + \beta)$ from the formula

$$\sin(\alpha - \beta) = \sin \alpha \cos \beta - \cos \alpha \sin \beta.$$

2. *Show that $\tan(\alpha - \beta) = \dfrac{\tan \alpha - \tan \beta}{1 + \tan \alpha \tan \beta}$, given the equation for $\tan(\alpha + \beta)$.

3. Prove by means of equations in this section that $\tan\left(\dfrac{\pi}{2} - \alpha\right) = \cot \alpha$.

4. Derive an equation for $\cot(\alpha + \beta)$ in terms of $\cot \alpha$ and $\cot \beta$.

Find the exact value of each of the following by means of the addition and subtraction identities.

5. $\sin \dfrac{5\pi}{12}$ 6. $\tan \dfrac{7\pi}{12}$ 7. $\cos \dfrac{\pi}{12}$

8. $\cos \dfrac{11\pi}{12}$ $\left(\text{Hint: } \dfrac{11\pi}{12} = \dfrac{\pi}{2} + \dfrac{5\pi}{12}. \text{ Use result of problem 5.}\right)$

Find, to the nearest degree, the greater angle between the two lines

9. whose slopes are $m_1 = \dfrac{2}{3}$ and $m_2 = 2$.

10. whose equations are $y = \dfrac{1}{3}x + 2$ and $3x - y - 6 = 0$.

Verify each of the following identities.

11. $\sin\left(\alpha + \dfrac{\pi}{4}\right) = \dfrac{1}{2}\sqrt{2}(\sin \alpha + \cos \alpha)$

12. $\tan\left(\dfrac{\pi}{4} + \alpha\right) = \dfrac{\cos \alpha + \sin \alpha}{\cos \alpha - \sin \alpha}$

13. $\tan \alpha + \tan \beta = \dfrac{\sin(\alpha + \beta)}{\cos \alpha \cos \beta}$

14. $\dfrac{\cos(x - y) - \cos(x + y)}{\cos(x - y) + \cos(x + y)} = \tan x \tan y$

15. a. $\sin\left(\dfrac{\pi}{2} + \alpha\right) = \cos \alpha$

---

* M. F. Spotts, *Design of Machine Elements*, fourth edition (Englewood Cliffs, N. J.: Prentice-Hall, 1971).

b. $\sin\left(\dfrac{3\pi}{2} - \alpha\right) = -\cos \alpha$

c. $\sin\left(\dfrac{3\pi}{2} + \alpha\right) = -\cos \alpha$

16. a. $\cos\left(\dfrac{\pi}{2} + \alpha\right) = -\sin \alpha$

b. $\cos\left(\dfrac{3\pi}{2} - \alpha\right) = -\sin \alpha$

c. $\cos\left(\dfrac{3\pi}{2} + \alpha\right) = \sin \alpha$

17. You will note from the results of problems 15 and 16 above that if $\alpha = \alpha'$ and $0 < \alpha' < \pi/2$, α' is a reference angle with respect to the y axis. Use the equations in problems 15 and 16 to show that

a. $\tan\left(\dfrac{\pi}{2} + \alpha\right) = -\cot \alpha$

b. $\tan\left(\dfrac{3\pi}{2} - \alpha\right) = \cot \alpha$

c. $\tan\left(\dfrac{3\pi}{2} + \alpha\right) = -\cot \alpha$

18. Why is it not possible to use equations (7) and (8) of Section 3.2 to derive the formulas of problem 17?

19. If $\cos \alpha = -3/5$ and $\pi/2 < \alpha < \pi$ while $\tan \beta = 5/12$ and $\pi < \beta < 3\pi/2$, find the exact value of

a. $\sin \alpha$ b. $\tan \alpha$
c. $\tan (\alpha + \beta)$ d. $\sin \beta$
e. $\cos \beta$ f. $\cos (\alpha + \beta)$
g. $\sin (\alpha - \beta)$
h. $\tan 2\alpha$. *Hint:* $\tan 2\alpha = \tan (\alpha + \alpha)$

20. If $\tan \alpha = 7/24$ and $0 < \alpha < \pi/2$ while $\sin \beta = \sqrt{39}/8$ and $\pi/2 < \beta < \pi$, find the exact value of

a. $\cos \beta$ b. $\tan \beta$
c. $\tan (\alpha - \beta)$ d. $\sin \alpha$
e. $\cos \alpha$ f. $\sin (\alpha + \beta)$
g. $\cos (\alpha - \beta)$
h. $\cos 2\alpha$. *Hint:* $\cos 2\alpha = \cos (\alpha + \alpha)$.

To show that an equation is not *an identity*

we can use a COUNTEREXAMPLE *which shows that the statement is not true for all values of the variable for which both members are defined. For example,* $\sin 2\alpha \neq 2 \sin \alpha$, *since upon substituting* $\dfrac{\pi}{6}$ *for* α, $\sin\left(2 \cdot \dfrac{\pi}{6}\right) = \sin \dfrac{\pi}{3} = \dfrac{1}{2}\sqrt{3}$, *while* $2 \sin \dfrac{\pi}{6} = 2 \cdot \dfrac{1}{2} = 1$, *and* $\dfrac{1}{2}\sqrt{3} \neq 1$.

21. Prove that $\sin 2\alpha = 2 \sin \alpha \cos \alpha$. *Hint:* $\sin 2\alpha = \sin (\alpha + \alpha)$.

22. a. Prove by a counterexample that $\tan 2\alpha \neq 2 \tan \alpha$

b. Prove that $\tan 2\alpha = \dfrac{2 \tan \alpha}{1 - \tan^2 \alpha}$

23. a. Prove by a counterexample that $\cos 2\alpha \neq 2 \cos \alpha$ b. Prove that $\cos 2\alpha = \cos^2 \alpha - \sin^2 \alpha$

24. a. Write an expression for $\cos 2\alpha$ in terms of $\cos \alpha$ alone. *Hint:* $\sin^2 \alpha + \cos^2 \alpha = 1$

b. Write an expression for $\cos 2\alpha$ in terms of $\sin \alpha$ alone.

25. Show that $\sin (\alpha + \beta) + \sin (\alpha - \beta) = 2 \sin \alpha \cos \beta$

26. Find the value of

$$\sin (\alpha + \beta) - \sin (\alpha - \beta)$$

in terms of sines or cosines of α and β.

27. Show that $\cos (\alpha + \beta) + \cos (\alpha - \beta) = 2 \cos \alpha \cos \beta$.

28. Find the value of

$$\cos (\alpha + \beta) - \cos (\alpha - \beta).$$

If we replace $\alpha + \beta$ *with* θ *and* $\alpha - \beta$ *with* ϕ *in the equations in problems 25–28 above we can see that*

$$\sin \theta + \sin \phi = 2 \sin \tfrac{1}{2}(\theta + \phi)$$
$$\times \cos \tfrac{1}{2}(\theta - \phi) \qquad (9)$$

$$\sin \theta - \sin \phi = 2 \cos \tfrac{1}{2}(\theta + \phi)$$
$$\times \sin \tfrac{1}{2}(\theta - \phi) \qquad (10)$$

$$\cos \theta + \cos \phi = 2 \cos \tfrac{1}{2}(\theta + \phi)$$
$$\times \cos \tfrac{1}{2}(\theta - \phi) \qquad (11)$$

$$\cos \theta - \cos \phi = -2 \sin \tfrac{1}{2}(\theta + \phi)$$
$$\times \sin \tfrac{1}{2}(\theta - \phi) \quad (12)$$

These are sometimes called SUMMATION FORMULAS *to distinguish them from the addition and subtraction identities.*

29. Verify that if $\alpha + \beta = \theta$ and $\alpha - \beta = \phi$, then $\alpha = \dfrac{1}{2}(\theta + \phi)$ and $\beta = \dfrac{1}{2}(\theta - \phi)$.

30. Examine the preceding exercises and find expressions for each of the following in terms of $\alpha + \beta$ and $\alpha - \beta$:
   a. $\sin \alpha \cos \beta$
   b. $\cos \alpha \sin \beta$
   c. $\cos \alpha \cos \beta$
   d. $\sin \alpha \sin \beta$

31. Express each of the following as a product:
   a. $\sin 4\alpha + \sin 2\alpha$
   b. $\cos \dfrac{\pi}{6} + \cos \dfrac{5\pi}{6}$
   c. $\sin 6x - \sin 4x$
   d. $\cos \dfrac{\pi}{3} - \cos \dfrac{4\pi}{3}$

32. Express each of the following as a sum or difference:
   a. $\sin 60° \cos 30°$
   b. $\cos 3x \cos x$
   c. $\sin 60° \sin 20°$

3.3 DOUBLE AND HALF-ANGLE FORMULAS

You probably encountered the DOUBLE-ANGLE IDENTITIES in the exercises of the previous section. These are the equations

$$\sin 2\alpha = 2 \sin \alpha \cos \alpha \qquad (1)$$

$$\cos 2\alpha = \cos^2 \alpha - \sin^2 \alpha \qquad (2)$$

$$= 1 - 2 \sin^2 \alpha \qquad (3)$$

$$= 2 \cos^2 \alpha - 1 \qquad (4)$$

$$\tan 2\alpha = \frac{2 \tan \alpha}{1 - \tan^2 \alpha} \qquad (5)$$

which can be derived from the addition identities by replacing β with α, as in $\sin 2\alpha = \sin(\alpha + \alpha) = \sin \alpha \cos \alpha + \cos \alpha \sin \alpha = 2 \sin \alpha \cos \alpha$. Equations (3) and (4) are derived from (2) by means of the Pythagorean identity $\sin^2 \alpha + \cos^2 \alpha = 1$. Since $\cos^2 \alpha = 1 - \sin^2 \alpha$, $\cos 2\alpha = \cos^2 \alpha - \sin^2 \alpha = (1 - \sin^2 \alpha) - \sin^2 \alpha = 1 - 2 \sin^2 \alpha$.

The above equations will appear in many places later in this text and in the calculus, but one of the main points is that they are true for any angle α, and that 2α is "twice α" whatever it may be. In other words, if $\alpha = 3\theta$, $2\alpha = 6\theta$, and $\sin 6\theta = 2 \sin 3\theta \cos 3\theta$; if $\alpha = 5x$, $\cos 10x = 1 - 2 \sin^2 5x$; if $\alpha = \tfrac{1}{2}t$, $2 \cos^2(\tfrac{1}{2}t) - 1 = \cos(2 \cdot \tfrac{1}{2}t) = \cos t$, the latter by applying equation (4) from right to left.

Again, different forms of the identities can be useful. If we solve equation (3) for $\sin^2 \alpha$, we obtain

$$\cos 2\alpha = 1 - 2 \sin^2 \alpha$$

or
$$2\sin^2\alpha = 1 - \cos 2\alpha$$

from which
$$\sin^2\alpha = \frac{1 - \cos 2\alpha}{2} \tag{6}$$

This equation is useful in itself, but from it we can derive one of the HALF-ANGLE IDENTITIES since α, on the left, is half of 2α, the angle on the right. Since equation (6) is true for all values of α

$$\sin^2 t = \frac{1 - \cos 2t}{2}$$

and if we replace t with $\tfrac{1}{2}\alpha$, then

$$\sin^2(\tfrac{1}{2}\alpha) = \frac{1 - \cos(2 \cdot \tfrac{1}{2}\alpha)}{2} = \frac{1 - \cos\alpha}{2}$$

$$\sin \tfrac{1}{2}\alpha = \pm\sqrt{\frac{1 - \cos\alpha}{2}} \tag{7}$$

By means of equation (4) we can derive

$$\cos \tfrac{1}{2}\alpha = \pm\sqrt{\frac{1 + \cos\alpha}{2}} \tag{8}$$

The ambiguous sign in equations (7) and (8) is determined by the quadrant in which $\tfrac{1}{2}\alpha$ terminates. For example,

$$\sin 150° = +\sqrt{\frac{1 - \cos 300°}{2}}$$

but
$$\cos 150° = -\sqrt{\frac{1 + \cos 300°}{2}}$$

since 150° terminates in Quadrant II. Sometimes it is a bit more difficult to determine the appropriate sign.

EXAMPLE 1   Given that $\sin\alpha = -\tfrac{3}{5}$ and $270° < \alpha < 360°$, find $\cos\tfrac{1}{2}\alpha$.

SOLUTION $\cos\tfrac{1}{2}\alpha = \pm\sqrt{\dfrac{1 + \cos\alpha}{2}}$, so we need $\cos\alpha$. Using the Pythagorean identity $\cos^2\alpha = 1 - \sin^2\alpha$, we obtain

$$\cos^2\alpha = 1 - \left(-\tfrac{3}{5}\right)^2 = 1 - \tfrac{9}{25} = \tfrac{16}{25}$$

$$\cos\alpha = \pm\tfrac{4}{5}$$

We choose $\cos \alpha = \dfrac{4}{5}$ since α is in Quadrant IV. The angle $\frac{1}{2}\alpha$, on the other hand, terminates in Quadrant II, since if $270° < \alpha < 360°$, $135° < \frac{1}{2}\alpha < 180°$. Thus $\cos \frac{1}{2}\alpha$ is negative, and

$$\cos \frac{1}{2}\alpha = -\sqrt{\frac{1 + \frac{4}{5}}{2}} = -\sqrt{\frac{9}{10}}$$

or $\cos \dfrac{1}{2}\alpha = -\dfrac{3}{\sqrt{10}} = -\dfrac{3}{10}\sqrt{10}$, depending on how much you may want to simplify the radical expression.

In order to find $\tan \frac{1}{2}\alpha$ we can first determine $\sin \frac{1}{2}\alpha$ and $\cos \frac{1}{2}\alpha$, then $\tan \frac{1}{2}\alpha = \dfrac{\sin \frac{1}{2}\alpha}{\cos \frac{1}{2}\alpha}$. A more direct approach would be to use the formula

$$\tan \frac{1}{2}\alpha = \pm \sqrt{\frac{1 - \cos \alpha}{1 + \cos \alpha}}$$

which can be derived as follows:

$$\sin^2 \alpha = \frac{1}{2}(1 - \cos 2\alpha) \text{ from equation (6)}$$

and

$$\cos^2 \alpha = \frac{1}{2}(1 + \cos 2\alpha) \text{ from equation (4)}$$

thus

$$\tan^2 \alpha = \frac{\sin^2 \alpha}{\cos^2 \alpha} = \frac{1 - \cos 2\alpha}{1 + \cos 2\alpha}$$

If α is replaced by $\frac{1}{2}\alpha$,

$$\tan^2 \frac{1}{2}\alpha = \frac{1 - \cos \alpha}{1 + \cos \alpha}$$

and

$$\tan \frac{1}{2}\alpha = \pm \sqrt{\frac{1 - \cos \alpha}{1 + \cos \alpha}} \tag{9}$$

Another equation for $\tan \frac{1}{2}\alpha$, which eliminates the need to decide on the appropriate sign, is

$$\tan \frac{1}{2}\alpha = \frac{1 - \cos \alpha}{\sin \alpha} \tag{10}$$

which can be derived by the following method:

$$\tan \tfrac{1}{2}\alpha = \frac{\sin \tfrac{1}{2}\alpha}{\cos \tfrac{1}{2}\alpha} \cdot \frac{\sin \tfrac{1}{2}\alpha}{\sin \tfrac{1}{2}\alpha} = \frac{\sin^2 \tfrac{1}{2}\alpha}{\sin \tfrac{1}{2}\alpha \cos \tfrac{1}{2}\alpha}$$

Replacing α with $\tfrac{1}{2}\alpha$ in equation (1), we find

$$\sin \alpha = 2 \sin \tfrac{1}{2}\alpha \cos \tfrac{1}{2}\alpha$$

$$\tfrac{1}{2} \sin \alpha = \sin \tfrac{1}{2}\alpha \cos \tfrac{1}{2}\alpha$$

Substituting this in the denominator and using equation (6) in the numerator,

$$\tan \tfrac{1}{2}\alpha = \frac{\tfrac{1}{2}(1 - \cos \alpha)}{\tfrac{1}{2} \sin \alpha} = \frac{1 - \cos \alpha}{\sin \alpha}$$

The identities derived in this and previous sections can be used to prove (and derive) a wide range of additional identities.

EXAMPLE 2   Prove that $\sin 2\alpha = \dfrac{2 \tan \alpha}{1 + \tan^2 \alpha}$ is an identity.

SOLUTION

$$\text{RM} = \frac{2 \tan \alpha}{\sec^2 \alpha} = 2 \tan \alpha \cdot \frac{1}{\sec^2 \alpha}$$

$$= 2 \cdot \frac{\sin \alpha}{\cos \alpha} \cdot \cos^2 \alpha$$

$$= 2 \sin \alpha \cos \alpha = \sin 2\alpha$$

EXAMPLE 3   Prove that $\tan \tfrac{1}{2}x = \csc x - \cot x$.

SOLUTION   Using equation (10), we can write

$$\text{LM} = \frac{1 - \cos x}{\sin x} = \frac{1}{\sin x} - \frac{\cos x}{\sin x} = \csc x - \cot x$$

Among the applications of the half-angle formulas we find the problem of determining the exact value of the trigonometric functions for certain angles.

EXAMPLE 4   Find the exact value of $\cos 75°$.

SOLUTION   Since $75°$ is half of $150°$

$$\cos 75° = \sqrt{\frac{1 + \cos 150°}{2}} = \sqrt{\frac{1 + (-\tfrac{1}{2}\sqrt{3})}{2}}$$

$$= \sqrt{\frac{2 - \sqrt{3}}{4}} = \tfrac{1}{2}\sqrt{2 - \sqrt{3}}$$

3.3 DOUBLE AND HALF-ANGLE FORMULAS

A somewhat complicated application of the identities in this chapter can be found in any text on analytic geometry. It involves simplifying equations of the form $Ax^2 + Bxy + Cy^2 = F$ by rotating the axis of the graph. The derivation of the transformation equations depends upon the addition identities of Section 3.2, but as an example we will simply show the half-angle identities involved.

It can be proved that $\cot 2\theta = \dfrac{A - C}{B}$—where A, B, and C are the coefficients of x^2, xy, and y^2 in the above equation—will give the angle θ required to rotate the axis and eliminate the xy term of the equation. Thus if we have $25x^2 - 24xy + 32y^2 = 64$,

$$\cot 2\theta = \frac{25 - 32}{-24} = \frac{7}{24}$$

If we assume that $0° < 2\theta < 180°$, then 2θ must terminate in Quadrant I since $\cot 2\theta > 0$, and we can find

$$\cos 2\theta = \frac{7}{25}$$

Now we can use the half-angle relations

$$\sin \theta = \sqrt{\frac{1 - \cos 2\theta}{2}} \quad \text{and} \quad \cos \theta = \sqrt{\frac{1 + \cos 2\theta}{2}}$$

to determine

$$\sin \theta = \sqrt{\frac{1 - \frac{7}{25}}{2}} = \frac{3}{5} \quad \text{and} \quad \cos \theta = \sqrt{\frac{1 + \frac{7}{25}}{2}} = \frac{4}{5}$$

from which we obtain

$$\tan \theta = \frac{3}{4}$$

The analytic geometry text will show how to obtain the equations

$$x = u \cos \theta - v \sin \theta \quad \text{and} \quad y = u \sin \theta + v \cos \theta$$

which can now be substituted into the equation $25x^2 - 24xy + 32y^2 = 64$ to obtain

$$16u^2 + 41v^2 = 64$$

We recognize the latter equation as that of an ellipse, and it is relatively easy to graph. In Figure 3–3 we see that the u axis has been rotated through the angle θ.

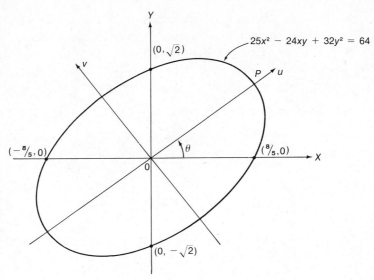

FIGURE 3-3

This is easy to do since $\tan \theta = 3/4$ means that the slope of the line OP is $3/4$ (relative to the x, y system). We can check to see that the points which are x and y intercepts are correct by substituting zero first for y, then for x, in the equation $25x^2 - 24xy + 32y^2 = 64$. Finding other points would be rather difficult, however, since each substitution would then require the solution of a quadratic equation. If $x = 1$, for example, the equation to solve is $25 - 24y + 32y^2 = 64$, or $32y^2 - 24y - 39 = 0$, and this would give us only two more points on the graph. A good sketch would require many more, particularly if we did not know the location of the axes of the ellipse.

EXERCISES

Use the double-angle or half-angle identities to write an equivalent expression in terms of 4θ.

1. $\sin 8\theta$
2. $\cos 8\theta$
3. $\dfrac{2 \tan 2\theta}{1 - \tan^2 2\theta}$
4. $\sin 2\theta \cos 2\theta$

EXAMPLE 5   a. $\pm \sqrt{\dfrac{1 - \cos 8\theta}{2}}$

b. $\tan 8\theta$

5. $2 \cos^2 2\theta - 1$
6. $\sin^2 2\theta - \cos^2 2\theta$
7. $\sqrt{\dfrac{1 + \cos 8\theta}{2}}$
8. $\dfrac{1 - \cos 8\theta}{\sin 8\theta}$

SOLUTION   a. Since 4θ is half of 8θ,

$$\pm \sqrt{\dfrac{1 - \cos 8\theta}{2}} = \sin 4\theta.$$

b. Since 8θ is twice 4θ,

$$\tan 8\theta = \dfrac{2 \tan 4\theta}{1 - \tan^2 4\theta}.$$

9. $\sin^2 2\theta$
10. $\tan^2 2\theta$

Use the half-angle formulas to find the exact value of

11. $\sin 15°$
12. $\cos 15°$
13. $\tan 15°$
14. $\tan 75°$

15. cot 75°
16. sin 75°
17. cos 150°
18. sin 285°

Given that $\cos \alpha = -3/5$ *and* $180° < \alpha < 270°$, *find the exact value of each of the following (show the equations used):*

19. $\sin 2\alpha$
20. $\cos 2\alpha$
21. $\tan 2\alpha$
22. $\cot 2\alpha$
23. $\cos \frac{1}{2}\alpha$
24. $\tan \frac{1}{2}\alpha$
25. $\sin 3\alpha$ *Hint:* $\sin 3\alpha = \sin(2\alpha + \alpha)$
26. $\cos 3\alpha$ *Hint:* $\cos 3\alpha = \cos(\alpha + 2\alpha)$

Prove that each of the following is an identity:

27. $(\cos \alpha - \sin \alpha)^2 = 1 - \sin 2\alpha$
28. $(\sin \alpha + \cos \alpha)^2 = 1 + \sin 2\alpha$
29. $\cos 2\alpha = \dfrac{1 - \tan^2 \alpha}{\sec^2 \alpha}$
30. $\sec^2 \dfrac{1}{2} u = \dfrac{2}{1 + \cos u}$
31. $\tan \dfrac{1}{2} \alpha = \dfrac{\sin \alpha}{1 + \cos \alpha}$
32. $\cot \dfrac{1}{2} x = \dfrac{1 + \cos x}{\sin x}$
33. $\sin 3\theta = 3 \sin \theta - 4 \sin^3 \theta$
34. $\cos 3\theta = 4 \cos^3 \theta - 3 \cos \theta$
35. $\cot^2 \alpha = \dfrac{\cos 2\alpha}{\sin^2 \alpha} + 1$
36. $1 + \tan^2 \alpha = \dfrac{2}{1 + \cos 2\alpha}$
37. $\cot 2\alpha = \dfrac{\csc \alpha - 2 \sin \alpha}{2 \cos \alpha}$
38. $\cot 2\alpha = \dfrac{\cot^2 \alpha - 1}{2 \cot \alpha}$
39. $\csc 2\alpha = \dfrac{\sec \alpha \csc \alpha}{2}$
40. $\sec 2\alpha = \dfrac{\sec^2 \alpha}{1 - \tan^2 \alpha}$
41. $\csc 2\alpha = \dfrac{1}{2}(\cot \alpha + \tan \alpha)$
42. $\tan 2u - \sec 2u = \dfrac{\sin u - \cos u}{\sin u + \cos u}$
43. $\dfrac{\cos 2t}{\sin t} + \dfrac{\sin 2t}{\cos t} = \csc t$
44. $\dfrac{\sin 2t}{\sin t} = \sec t + \dfrac{\cos 2t}{\cos t}$
45. $\dfrac{2 \sin 2u}{1 - \cos^2 2u} = \sec u \csc u$
46. $\dfrac{2 \sin^2 \alpha}{\sin 2\alpha} + \cot \alpha = \sec \alpha \csc \alpha$

47. The equation $5x^2 - 4xy + 8y^2 = 36$ can be transformed into the equation $4u^2 + 9v^2 = 36$ by rotating the axes through an angle θ determined by using $\cot 2\theta = (A - C)/B$ as shown in the last example of this section.

   a. Find $\sin \theta$, $\cos \theta$, and $\tan \theta$.
   b. Draw the u, v coordinate system with the u axis passing through the origin of the x, y system and having slope $\tan \theta$.
   c. Find y when $x = 0$ and x when $y = 0$ and plot the four points which result.
   d. Complete the graph of $4u^2 + 9v^2 = 36$ on the u, v coordinate axes.

3.4 CONDITIONAL EQUATIONS

Up to this point in this chapter we have been concerned with trigonometric identities, or equations which are true for all values of the variable for which each function in the equation is defined. There are times, however, when it will be necessary to solve a CONDITIONAL TRIGONOMETRIC EQUATION, or one for which some, or many, defined values of the function do not satisfy the equation.

These equations will be written in terms of one or more trigonometric functions, for example,

$$\cos x + 1 = 0 \qquad \cos^2 x + \cos x = 0 \qquad 2\cos^2 x + \cos x - 1 = 0$$

$$\cos x - \sec x = 0 \qquad \cos 2x + \cos x = 0$$

If we let $u = \cos x$, the first three of the above equations become $u + 1 = 0$, $u^2 + u = 0$, and $2u^2 + u - 1 = 0$, which are algebraic equations we would find easy to solve. We can transform the last two equations by writing equivalent equations and simplifying by means of the identities of the previous sections. We can change $\cos x - \sec x = 0$, for example, to $\cos x - 1/(\cos x) = 0$ or $\cos^2 x - 1 = 0$, which assumes the form $u^2 - 1 = 0$. To solve $\cos 2x + \cos x = 0$ we replace $\cos 2x$ with $2\cos^2 x - 1$, using equation (4) of the previous section, so that $2\cos^2 x - 1 + \cos x = 0$, which is really the same as one of the equations which precedes it in our list of examples.

The complete solution of the trigonometric equation requires us to list all of the angles or numbers which satisfy it. To solve $\cos x + 1 = 0$, we add -1 to both sides of the equation to obtain $\cos x = -1$. Had we been solving the equation $u + 1 = 0$, finding $u = -1$ would be the end of it since -1 satisfies the equation, that is, $(-1) + 1 = 0$. With $\cos x + 1 = 0$, however, finding $\cos x = -1$ is just a plateau. Now we must determine for which values of x it is true that $\cos x = -1$, but we have done this before, in Chapter Two. If $\cos x = -1$, $x = 180° + k \cdot 360°$ or $x = \pi + k \cdot 2\pi$. Thus the solution set for $\cos x + 1 = 0$ is $\{x \mid x = 180° + k \cdot 360°\}$ or $\{x \mid x = \pi + k \cdot 2\pi\}$. It is usually awkward or difficult to check all possible solutions, so one or two values will generally suffice. We can write $\cos 180° + 1 = (-1) + 1 = 0$ and feel reasonably comfortable that the remaining values will also satisfy the equation.

In order to solve $\cos^2 x + \cos x = 0$, we factor to obtain $\cos x (\cos x + 1) = 0$. Since the product is 0, either factor can equal 0, so $\cos x = 0$ or $\cos x + 1 = 0$. The solutions for $\cos x = 0$ are $x = 90° + k \cdot 180°$ or $x = \dfrac{\pi}{2} + k \cdot \pi$, while the equation $\cos x + 1 = 0$ is identical to that in the previous paragraph. Thus the solution set is $\{x \mid x = 90° + k \cdot 180°\} \cup \{x \mid x = 180° + k \cdot 360°\}$, which means the set of numbers in either the first set or the second set. The symbol $\cup$ stands for *union*. Checking with $x = 90°$, $\cos^2 90° + \cos 90° = (0)^2 + 0 = 0$, and with $x = 180°$, $\cos^2 180° + \cos 180° = (-1)^2 + (-1) = 1 + (-1) = 0$.

We usually would know whether we want to work with degree or radian measure, of course, and may even specify a part of the domain as a replacement set. To solve $2\cos^2 x + \cos x - 1 = 0$ over $0° \leq x < 360°$, for example, we first factor to obtain $(2\cos x - 1)(\cos x + 1) = 0$. Then $2\cos x - 1 = 0$ or $\cos x + 1 = 0$. The first equation reduces to $2\cos x = 1$ or $\cos x = \frac{1}{2}$, and its solution set is

3.4 CONDITIONAL EQUATIONS

$\{x \mid x = 60° + k \cdot 360°\} \cup \{x \mid x = 300° + k \cdot 360°\}$. The second has as its solution set $\{x \mid x = 180° + k \cdot 360°\}$. The values in the required interval are in the set $\{60°, 180°, 300°\}$. Checking $60°$ we find

$$2\cos^2 60° + \cos 60° - 1 = 2\left(\frac{1}{2}\right)^2 + \frac{1}{2} - 1$$

$$= 2 \cdot \frac{1}{4} + \frac{1}{2} - 1 = \frac{1}{2} + \frac{1}{2} - 1 = 0$$

Each of the other two values will also satisfy the original equation.

There is probably no general program for the solution of all trigonometric equations available. Our strategy will be essentially as follows:

1. If the equation contains more than one trigonometric function, transform it so that there is only one function by means of identities, or factor it into a product equal to zero, so that each factor will contain only one function;

2. If the equation contains functions of more than one angle, such as $2x$ and x, use the double and half-angle formulas to reduce it to functions of the same angle.

EXAMPLE 1   Solve $\sin^2 x = \cos^2 x$ where $0 \leq x < 2\pi$.

SOLUTION   We can divide both sides of the equation by $\cos^2 x$ to obtain

$$\frac{\sin^2 x}{\cos^2 x} = 1$$

This is equivalent to $\tan^2 x = 1$, so $\tan x = 1$ or $\tan x = -1$. If $\tan x = 1$, then $x = \pi/4$ or $x = 5\pi/4$, and if $\tan x = -1$, then $x = 3\pi/4$ or $x = 7\pi/4$. Each of these will satisfy the given equation, so the solution set is $\left\{\dfrac{\pi}{4}, \dfrac{3\pi}{4}, \dfrac{5\pi}{4}, \dfrac{7\pi}{4}\right\}$.

As in algebra we must be particularly careful to check solutions when we have multiplied or divided by an expression containing the variable or when we have squared both members, since we may have introduced an extraneous root.

EXAMPLE 2   Solve $\dfrac{1 - \sin x}{\cos x} = \cos x$ where $0 \leq x < 2\pi$.

SOLUTION   Multiply both members by cos x to find

$$1 - \sin x = \cos^2 x$$

Since $\cos^2 x = 1 - \sin^2 x$, we can transform this to

$$1 - \sin x = 1 - \sin^2 x$$
$$\sin^2 x - \sin x = 0$$
$$\sin x (\sin x - 1) = 0$$
$$\sin x = 0 \quad \text{or} \quad \sin x - 1 = 0$$
$$\sin x = 0 \quad \text{or} \quad \sin x = 1$$

If $\sin x = 0$, $x = 0$ or $x = \pi$ in the given interval. If $\sin x = 1$, then $x = \pi/2$. Checking each of these by substitution, we find that replacing x by $\pi/2$ gives

$$\frac{1 - \sin \frac{\pi}{2}}{\cos \frac{\pi}{2}} = \cos \frac{\pi}{2}$$

but since $\cos (\pi/2) = 0$ we have a division by 0, which is undefined. Thus the solution set is $\{0, \pi\}$.

EXAMPLE 3   Solve $\sin x - \cos 2x = 2$ where $-\pi < x \leq \pi$.

SOLUTION   Since there is a choice, use equation (3) of the previous section to replace $\cos 2x$, so that

$$\sin x - (1 - 2 \sin^2 x) = 2$$
$$2 \sin^2 x + \sin x - 3 = 0$$
$$(2 \sin x + 3)(\sin x - 1) = 0$$
$$2 \sin x + 3 = 0 \quad \text{or} \quad \sin x - 1 = 0$$
$$\sin x = -\frac{3}{2} \qquad \sin x = 1$$

There are no solutions for the first equation since the range of the sine function is $-1 \leq \sin x \leq 1$. For the second, $x \in \left\{x \mid x = \frac{\pi}{2} + k \cdot 2\pi\right\}$, so $\left\{\frac{\pi}{2}\right\}$ is the solution set over the specified replacement set.

3.4 CONDITIONAL EQUATIONS

EXAMPLE 4 Solve $\cos x = \sin 2x$ for all possible values of x in radian measure.

SOLUTION Replace $\sin 2x$ by means of a double-angle formula to obtain

$$\cos x = 2 \sin x \cos x$$
$$\cos x - 2 \sin x \cos x = 0$$
$$\cos x (1 - 2 \sin x) = 0$$
$$\cos x = 0 \quad \text{or} \quad 1 - 2 \sin x = 0$$

The second equation reduces to

$$\sin x = \frac{1}{2}$$

The solutions for $\cos x = 0$ are in the set $\{x \mid x = k \cdot \pi\}$, while the solutions for the second equation are in the union of $\left\{x \mid x = \dfrac{\pi}{6} + k \cdot 2\pi\right\}$ and $\left\{x \mid x = \dfrac{5\pi}{6} + k \cdot 2\pi\right\}$. Thus all solutions are contained in

$$\{x \mid x = k \cdot \pi\} \cup \left\{x \mid x = \frac{\pi}{6} + k \cdot 2\pi\right\} \cup \left\{x \mid x = \frac{5\pi}{6} + k \cdot 2\pi\right\}.$$

The reader should beware of dividing both sides of the equation $\cos x = 2 \sin x \cos x$ by $\cos x$ to simplify it. The result would be the equation $1 = 2 \sin x$ which is not equivalent and a significant portion of the solution set would be lost.

EXAMPLE 5 Solve $2 \cos 3x + 1 = 0$ where $0 \leq x < 360°$.

SOLUTION First let $3x = \alpha$. Then $2 \cos \alpha + 1 = 0$, so $\cos \alpha = -\frac{1}{2}$, and the solution set for the latter is given by $\alpha = 240° + k \cdot 360°$ or $\alpha = 300° + k \cdot 360°$. Thus if we replace α by $3x$, we obtain in the first instance $3x = 240° + k \cdot 360°$, so that $x = 80° + k \cdot 120°$, which means $x = 80°$, $x = 200°$, and $x = 320°$ in the interval. Substituting in the second equation, $3x = 300° + k \cdot 360°$, then $x = 100° + k \cdot 120°$, from which we find $x = 100°$, $x = 220°$, and $x = 340°$ in the interval. The solution set is $\{80°, 100°, 200°, 220°, 320°, 340°\}$.

EXAMPLE 6 Solve the equation $\sqrt{\dfrac{1 - \cos x}{2}} = \dfrac{1}{2}$ where $0 \leq x < 2\pi$.

SOLUTION   We can square both sides of the equation to obtain $\dfrac{1-\cos x}{2} = \dfrac{1}{4}$, from which we derive

$$1 - \cos x = \dfrac{1}{2}$$

$$-\cos x = -\dfrac{1}{2}$$

$$\cos x = \dfrac{1}{2}$$

The solution set for the last equation is $\left\{\dfrac{\pi}{3}, \dfrac{5\pi}{3}\right\}$. Squaring both members did not introduce an extraneous root in this case.

Another solution would be to replace $\sqrt{\dfrac{1-\cos x}{2}}$ by $\sin \dfrac{1}{2}x$, so that

$$\sin \dfrac{1}{2}x = \dfrac{1}{2}$$

Then

$$\dfrac{1}{2}x = \dfrac{\pi}{6} + k \cdot 2\pi$$

so that

$$x = \dfrac{\pi}{3} + k \cdot 4\pi$$

or

$$\dfrac{1}{2}x = \dfrac{5\pi}{6} + k \cdot 2\pi$$

so that

$$x = \dfrac{5\pi}{3} + k \cdot 4\pi$$

and the same solution set would be obtained for the interval. Again some care should be exercised when making this replacement, since $\sin \dfrac{1}{2}x = \pm\sqrt{\dfrac{1-\cos x}{2}}$ and there may be some difficulty with the ambiguous sign.

EXERCISES

Solve each equation in problems 1–8 in the interval $0° \leq x < 360°$.

1. $2 \cos x + 1 = 0$
2. $1 + 2 \sin x = 0$
3. $4 \sin^2 x - 3 = 0$
4. $2 \cos^2 x - 1 = 0$
5. $\sin 3x = 1$
6. $\cos 4x + 1 = 0$
7. $\sqrt{3} \tan 2x + 1 = 0$
8. $\sqrt{3} \cot 3x - 1 = 0$

Solve each equation in problems 9–12 in the interval $0 \le x < 2\pi$.

9. $\sin 2x + \cos x = 0$
10. $\tan x - \sin 2x = 0$
11. $\cos^2 \frac{2}{3}x - 1 = 0$
12. $2 \sin^2 \frac{4}{3}x - 1 = 0$

Solve each equation in problems 13–16 if the replacement set is the set of all angles and their measure is in degrees.

13. $\sec^2 x - 2 \tan x = 0$
14. $2 \sin x + \cot x = \csc x$
15. $\sec x + \cos x = \sin x \tan x$
16. $\cos 2x + 2 \sin x \cos x = 0$

Solve problems 17–34 for all values of x in $0° \le x < 360°$.

17. $2 \sin^2 x - \sin x = 0$
18. $\tan^2 x + 2 \tan x = 0$
19. $\cos^2 x - 3 \cos x + 2 = 0$
20. $2 \sin^2 x - \sin x - 3 = 0$
21. $\tan^2 x + 2 \tan x - 3 = 0$
22. $\sec^2 x + 5 \sec x + 4 = 0$
23. $5 \sin x + 3 = 0$
24. $3 \sec x - 5 = 0$
25. $3 \cot^2 x - \cot x - 2 = 0$
26. $6 \cos^2 x - \cos x - 1 = 0$
27. $2 \cos x - \sec x + 1 = 0$
28. $5 \cot x + \tan x + 6 = 0$
29. $5 \sin^2 x - 2 \sin x - 1 = 0$. *Hint: Use the quadratic formula.*
30. $2 \sec^2 x - \sec x - 4 = 0$
31. $3 \cot^2 x + 2 \cot x - 2 = 0$
32. $\csc^2 x - 4 \csc x - 1 = 0$
33. $\cos^2 2x - 2 \cos 2x - 1 = 0$
34. $\tan 2x = 6 - 7 \cot 2x$

Solve problems 35–40 for all values of x in $-\frac{\pi}{2} \le x \le \frac{\pi}{2}$.

35. $\sin\left(x + \frac{\pi}{6}\right) = \frac{1}{2}$
36. $\tan\left(x + \frac{\pi}{3}\right) = 1$
37. $\sqrt{2} \cos\left(x + \frac{\pi}{3}\right) = 1$
38. $2 \sin\left(x - \frac{2\pi}{3}\right) + \sqrt{3} = 0$
39. $\tan\left(2x - \frac{\pi}{4}\right) + 1 = 0$
40. $\cos 3\left(x + \frac{\pi}{3}\right) = 0$

REVIEW EXERCISES

Prove that each of the following is an identity:

1. $\sin x \sec x = \tan x$
2. $\dfrac{1 - \cos 2x}{\sin 2x} = \tan x$
3. $\sec^2 x = \csc^2 x \tan^2 x$
4. $\sin 2x = \dfrac{2 \tan x}{1 + \tan^2 x}$
5. $\dfrac{\cos^2 x}{1 + \sin x} + \dfrac{\cos^2 x}{1 - \sin x} = 2$
6. $\sin 2x \tan 2x = \tan x (\sin 2x + \tan 2x)$

7. $\dfrac{1}{1 - \cos x} + \dfrac{1}{1 + \cos x} = 2 \cot x \csc x$

8. $\tan \dfrac{x}{2} + \cot \dfrac{x}{2} = 2 \csc x$

9. $\dfrac{1 + \sin x - \cos 2x}{\cos x + \sin 2x} = \tan x$

10. $\sin 4x = 4 \cos x (\sin x - 2 \sin^3 x)$

11. Given that $\cos x = 5/13$ when $270° < x < 360°$, and $\cot y = -15/8$ when $90° < y < 180°$, find the exact values of

    a. $\sin x$ b. $\cot x$

    c. $\sin y$ d. $\cos y$

    e. $\sin (x - y)$ f. $\tan 2x$

    g. $\sin \dfrac{x}{2}$ h. $\tan 3x$

12. Calculate the exact value of $\cos 75°$ by means of   a. an addition formula   b. a half-angle formula   c. Show that the results in parts a and b are equal.

13. Solve for x in the interval $0° \leq x < 360°$:

    a. $\sin 2x = \sin x$

    b. $\tan 3x + 1 = 0$

    c. $\cos^2 x - \sin^2 x = 1$

    d. $\cos 2x + \sin x = 0$

    e. $\cos 2x = 2 \cos x$

    f. $\tan x = \sec^2 x - 3$

CHAPTER FOUR

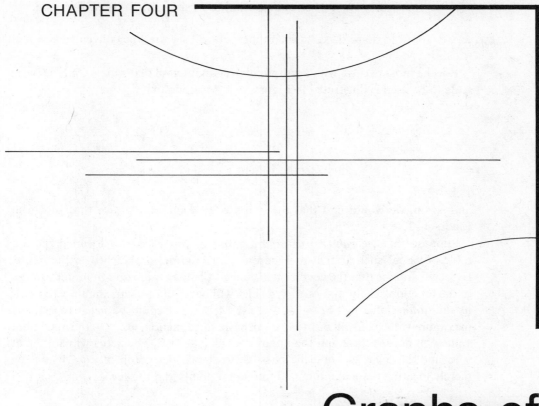

Graphs of Trigonometric Functions and Inverse Functions

4.1 GRAPHS OF $y = A \sin Bx$ AND $y = A \cos Bx$

At the end of Chapter Two we laboriously plotted the graphs of the six basic trigonometric functions. We are now ready to learn how to sketch these graphs rapidly, a skill necessary for future work in mathematics, engineering, and science. To do this, let us agree to a convenient scale which will facilitate the rapid sketching process.

You are familiar with the values of the trigonometric functions for 0, $\pi/6$, $\pi/4$, $\pi/3$, $\pi/2$, $2\pi/3$, and so on, and since these are all multiples of $\pi/12$, that is, $\pi/6 = 2 \cdot \pi/12$, $\pi/4 = 3 \cdot \pi/12$, $\pi/3 = 4 \cdot \pi/12$, and so on, we shall use a scale of 1 square $= \pi/12$ units on the horizontal and 5 squares $= 1$ unit on the vertical. This results in a slight distortion, but considering the stretching, squeezing, and shifting which we are about to do to these basic graphs, you will see that the convenience is well worth it.

For the most part we will be graphing over an interval of $0 \leq x \leq 2\pi$. The basic scale to be used is illustrated in Figure 4–1. Recalling that

$$\sin \frac{\pi}{6} = \frac{1}{2} = 0.50 \qquad \sin \frac{\pi}{4} = \frac{1}{2}\sqrt{2} \approx 0.71 \qquad \sin \frac{\pi}{3} = \frac{1}{2}\sqrt{3} \approx 0.87$$

$$\sin \frac{\pi}{2} = 1.00 \qquad \sin \frac{2\pi}{3} = \sin \frac{\pi}{3} \approx 0.87$$

and so on, we would find that the graph of $y = \sin x$ looks like that shown in Figure 4–2.

Suppose that we want to draw the graph of $y = 2 \sin x$. We can prepare a table of values such as that in Figure 4–3. The graph is sketched in the figure. Especially notice that the coefficient, 2, affects the maximum and minimum values of the function. The graph of $y = A \sin x$ will have $|A|$ as its maximum value and its minimum value will be $-|A|$. We use $|A|$ since A could be negative, but the maximum value will still be $|A|$. For the basic sine function, if $A > 0$ the maximum value will occur at $\pi/2$ and the minimum value at $3\pi/2$. The zeros will still be at 0, π, and 2π. Thus it would be possible to sketch the graph of $y = A \sin x$ by merely locating the maximum and minimum points and the zeros.

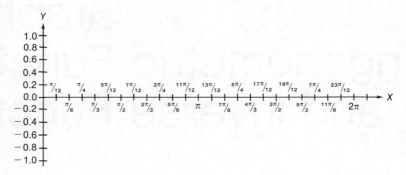

FIGURE 4–1

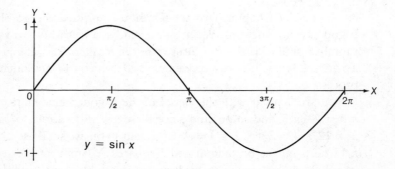

FIGURE 4–2

4.1 GRAPHS OF $y = A \sin Bx$ AND $y = A \cos Bx$

x	0	$\frac{\pi}{6}$	$\frac{\pi}{4}$	$\frac{\pi}{3}$	$\frac{\pi}{2}$	$\frac{5\pi}{6}$	π	$\frac{7\pi}{6}$	$\frac{3\pi}{2}$	$\frac{11\pi}{6}$	2π
$\sin x$	0	$\frac{1}{2}$	$\frac{1}{2}\sqrt{2}$	$\frac{1}{2}\sqrt{3}$	1	$\frac{1}{2}$	0	$-\frac{1}{2}$	-1	$-\frac{1}{2}$	0
$2 \sin x$	0	1.00	1.41	1.73	2.00	1.00	0	-1.00	-2.00	-1.00	0

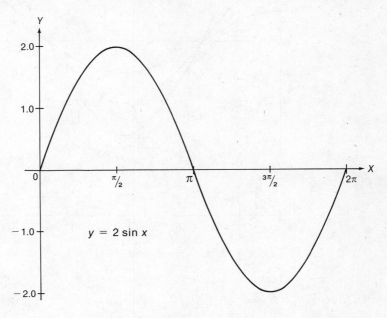

FIGURE 4–3

EXAMPLE 1 Sketch the graph of $y = \sqrt{2} \sin x$ where $0 \leq x \leq 2\pi$.

SOLUTION $A = \sqrt{2}$, so the maximum value is $\sqrt{2}$, the minimum value is $-\sqrt{2}$. These occur at $\pi/2$ and $3\pi/2$. We plot these points and the zeros, then sketch the graph, as shown in Figure 4–4.

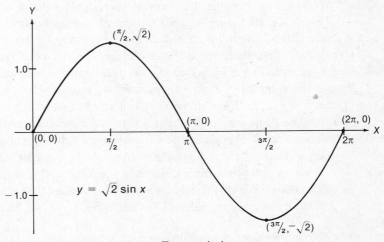

FIGURE 4–4

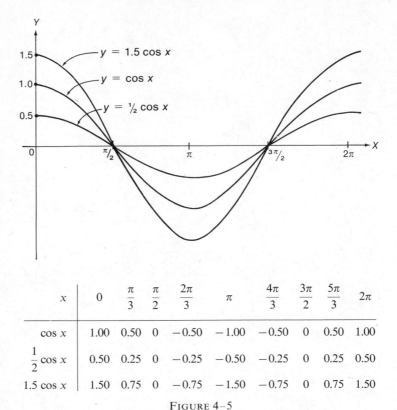

x	0	$\dfrac{\pi}{3}$	$\dfrac{\pi}{2}$	$\dfrac{2\pi}{3}$	π	$\dfrac{4\pi}{3}$	$\dfrac{3\pi}{2}$	$\dfrac{5\pi}{3}$	2π
$\cos x$	1.00	0.50	0	-0.50	-1.00	-0.50	0	0.50	1.00
$\dfrac{1}{2}\cos x$	0.50	0.25	0	-0.25	-0.50	-0.25	0	0.25	0.50
$1.5\cos x$	1.50	0.75	0	-0.75	-1.50	-0.75	0	0.75	1.50

FIGURE 4–5

The graphs of $y = \cos x$, $y = \frac{1}{2}\cos x$, and $y = 1.5\cos x$ are shown in Figure 4–5, along with the plotting table used to make the sketch. Again we see that the maximum and minimum values are affected by the coefficient of $\cos x$. In equations of the form $y = A \sin x$ and $y = A \cos x$, the number $|A|$ is known as the AMPLITUDE of the function.*

Next let us examine the effect of varying B in the equations $y = \sin Bx$ and $y = \cos Bx$. In Figure 4–6 we see the plotting table and graph for $y = \sin 3x$. Note that in obtaining the values for the plotting table we multiplied each value of x by 3 first, then found the sine of $3x$. Had we been graphing $y = 3 \sin x$, at $\pi/6$ we would have found $\sin(\pi/6) = 1/2$, then $y = 3 \sin(\pi/6) = 3(\frac{1}{2}) = 1\frac{1}{2}$. To find the value of $y = \sin 3x$, however, we multiplied 3 by $\pi/6$ to obtain $\pi/2$, so that $y = \sin(3 \cdot \pi/6) = \sin(\pi/2) = 1$. The effect of multiplying x by 3, then, is to cause the basic sine curve to run through its values three times as fast. In the interval from 0 through 2π the basic curve goes through all of its successive values once, whereas in $y = \sin 3x$, it goes through these values three times. We say that $y = \sin 3x$ completes three CYCLES in the interval $0 \leq x \leq 2\pi$.

The reason for this is evident when we consider that the zeros of the function $y = \sin x$ occur at intervals of $k \cdot \pi$. The zeros of $y = \sin 3x$ occur when $3x = k \cdot \pi$,

---

* Strictly speaking, the amplitude of a sine or cosine curve is the absolute value of one-half the difference between the maximum and minimum values of the function.

4.1 GRAPHS OF $y = A \sin Bx$ AND $y = A \cos Bx$

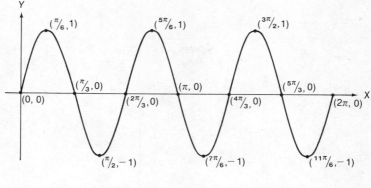

x	0	$\dfrac{\pi}{12}$	$\dfrac{\pi}{6}$	$\dfrac{\pi}{4}$	$\dfrac{\pi}{3}$	$\dfrac{5\pi}{12}$	$\dfrac{\pi}{2}$	$\dfrac{7\pi}{12}$	$\dfrac{2\pi}{3}$
$3x$	0	$\dfrac{\pi}{4}$	$\dfrac{\pi}{2}$	$\dfrac{3\pi}{4}$	π	$\dfrac{5\pi}{4}$	$\dfrac{3\pi}{2}$	$\dfrac{7\pi}{4}$	2π
$\sin 3x$	0	0.7	1.0	0.7	0	-0.7	-1.0	-0.7	0
x	$\dfrac{3\pi}{4}$	$\dfrac{5\pi}{6}$	π	$\dfrac{7\pi}{6}$	$\dfrac{4\pi}{3}$	$\dfrac{3\pi}{2}$	$\dfrac{5\pi}{3}$	$\dfrac{11\pi}{6}$	2π
$3x$	$\dfrac{9\pi}{4}$	$\dfrac{5\pi}{2}$	3π	$\dfrac{7\pi}{2}$	4π	$\dfrac{9\pi}{2}$	5π	$\dfrac{11\pi}{2}$	6π
$\sin 3x$	0.7	1.0	0	-1.0	0	1.0	0	-1.0	0

FIGURE 4–6

or when $x = k \cdot \pi/3$ (refer to Figure 4–6). The zeros of $y = \sin Bx$ occur when $Bx = k \cdot \pi$, or when $x = k \cdot \pi/B$. The period of $y = \sin x$ is 2π, as discussed in Section 2.4, and the period of $y = \sin Bx$, $(B > 0)$, is $2\pi/B$. The period of $y = \cos x$, $(B > 0)$, is also $2\pi/B$.

EXAMPLE 2   Draw the graph of $y = \frac{1}{2} \cos 2x$ where $0 \leq x \leq 2\pi$.

SOLUTION   Just once more let us prepare a plotting table. After this we will be able to sketch any curve of the form $y = A \sin Bx$ or $y = A \cos Bx$ without such a table. The graph and table are shown in Figure 4–7.

In Example 2 we found that the period of the graph is $2\pi/2 = \pi$, and that the graph completes two cycles in the interval $0 \leq x \leq 2\pi$. The amplitude of the graph is 1/2. The maximum values of the basic cosine curve occur when $x = k \cdot 2\pi$, so the maximum values of the curve $y = \frac{1}{2} \cos 2x$ occur when $2x = k \cdot 2\pi$, or when $x = k \cdot \pi$. The minimum values of the basic cosine curve occur when $x = \pi + k \cdot 2\pi$, so the minimum values of $y = \frac{1}{2} \cos 2x$ occur when $2x = \pi + k \cdot 2\pi$,

x	0	$\frac{\pi}{6}$	$\frac{\pi}{4}$	$\frac{\pi}{3}$	$\frac{\pi}{2}$	$\frac{2\pi}{3}$	$\frac{3\pi}{4}$	$\frac{5\pi}{6}$	π	$\frac{5\pi}{4}$	$\frac{3\pi}{2}$	$\frac{7\pi}{4}$	2π
$2x$	0	$\frac{\pi}{3}$	$\frac{\pi}{2}$	$\frac{2\pi}{3}$	π	$\frac{4\pi}{3}$	$\frac{3\pi}{2}$	$\frac{5\pi}{3}$	2π	$\frac{5\pi}{2}$	3π	$\frac{7\pi}{2}$	4π
$\cos 2x$	1	$\frac{1}{2}$	0	$-\frac{1}{2}$	-1	$-\frac{1}{2}$	0	$\frac{1}{2}$	1	0	-1	0	1
$\frac{1}{2}\cos 2x$	$\frac{1}{2}$	$\frac{1}{4}$	0	$-\frac{1}{4}$	$-\frac{1}{2}$	$-\frac{1}{4}$	0	$\frac{1}{4}$	$\frac{1}{2}$	0	$-\frac{1}{2}$	0	$\frac{1}{2}$

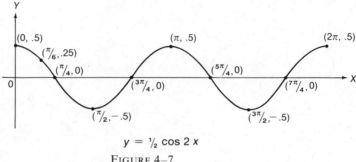

$y = \frac{1}{2}\cos 2x$

FIGURE 4–7

or when $x = \frac{\pi}{2} + k \cdot \pi$. The zeros of the basic cosine curve occur when $x = \frac{\pi}{2} + k \cdot \pi$, so the zeros of $y = \frac{1}{2}\cos 2x$ occur when $2x = \frac{\pi}{2} + k \cdot \pi$, or when $x = \frac{\pi}{4} + k \cdot \frac{\pi}{2}$.

Examining one cycle of the cosine curve we see that it has its maximum values at the endpoints of the cycle and a minimum value halfway between. There is a zero halfway between each endpoint and the minimum value.

EXAMPLE 3   Sketch the graph of one cycle of $y = \sqrt{3} \cos 3x$.

SOLUTION   Determine the period, $2\pi/B = 2\pi/3$, which is the length of the cycle. The amplitude is $\sqrt{3} \approx 1.73$. Locate the maximum values at $(0, \sqrt{3})$ and $(2\pi/3, \sqrt{3})$. The minimum value will be halfway between, at $(\pi/3, -\sqrt{3})$. The zeros will be halfway between these, that is, at $x = \pi/6$ and at $x = \pi/2$. Sketch the graph, using a smooth curve which is rounded out as experience has shown the cosine curve to be. Refer to Figure 4–8. Should we be asked to extend the graph, we see that the important points all occur at intervals of $\pi/6$, so it is easy to locate them and complete the sketch.

Finding the maximum or minimum points and the zeros of a graph can be considered simply the solutions of trigonometric equations. In the preceding example, for instance, the maximum values of $y = \sqrt{3} \cos 3x$ are $\sqrt{3}$, so if

4.1 GRAPHS OF $y = A \sin Bx$ AND $y = A \cos Bx$

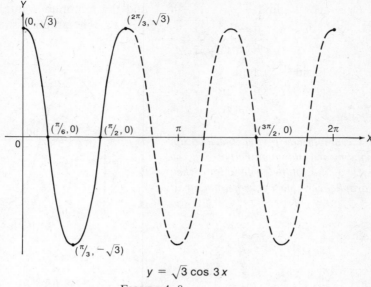

$$y = \sqrt{3} \cos 3x$$
FIGURE 4–8

$\sqrt{3} \cos 3x = \sqrt{3}$, $\cos 3x = 1$, and the solutions of this equation are $3x = k \cdot 2\pi$, or $x = k \cdot (2\pi/3)$. Thus the maximum values in the interval $0 \leq x \leq 2\pi$ are at 0, $\frac{2\pi}{3}$, $\frac{4\pi}{3}$, and 2π. The minimum values are $-\sqrt{3}$, so if $\sqrt{3} \cos 3x = -\sqrt{3}$, $\cos 3x = -1$, and the solutions of this equation are $3x = \pi + k \cdot 2\pi = \pi(1 + 2k)$, from which $x = \frac{\pi}{3}(1 + 2k)$, or odd multiples of $\frac{\pi}{3}$, that is, $\frac{\pi}{3}$, π, $\frac{5\pi}{3}$. The zeros occur when $\sqrt{3} \cos 3x = 0$, or $\cos 3x = 0$. The solutions for this equation are at $3x = \frac{\pi}{2} + k \cdot \pi = \frac{\pi}{2}(1 + 2k)$, or $x = \frac{\pi}{6}(1 + 2k)$, the odd multiples of $\frac{\pi}{6}$. Thus the zeros are at

$$\frac{\pi}{6}, \frac{\pi}{2}, \frac{5\pi}{6}, \frac{7\pi}{6}, \frac{3\pi}{2}, \frac{11\pi}{6}$$

EXERCISES

In preparing the problems in this chapter, I assumed that you would use graph paper with four or five squares per inch, and that the horizontal and vertical scales mentioned at the beginning of this section would be used except where noted.

Find the exact value of each of the following:

EXAMPLE 4 Given $f(x) = \sin 2x$, find
a. $f(0)$ b. $f(\pi/4)$ c. $f(5\pi/3)$

SOLUTION

a. $f(0) = \sin (2 \cdot 0) = \sin 0 = 0$
b. $f(\pi/4) = \sin (2 \cdot \pi/4) = \sin (\pi/2) = 1$
c. $f(5\pi/3) = \sin (2 \cdot 5\pi/3) = \sin (10\pi/3)$
$= \sin (4\pi/3 + 2\pi)$
$= \sin (4\pi/3) = \sin (\pi/3 + \pi)$
$= -\sin (\pi/3) = -\frac{1}{2}\sqrt{3}$

1. Given $f(x) = \cos 3x$, find
   a. $f(0)$
   b. $f(\pi/6)$
   c. $f(\pi/3)$
   d. $f(11\pi/6)$

2. Given $f(x) = 2 \sin (\frac{1}{2}x)$, find
   a. $f(\pi/3)$
   b. $f(3\pi/2)$
   c. $f(2\pi)$
   d. $f(5\pi/2)$

Determine the amplitude, period, the location of the maximum and minimum values and the zeros of each of the following functions, then draw the graph over one cycle beginning at $x = 0$.

3. $y = \sin x$
4. $y = \frac{4}{5} \sin x$
5. $y = \cos x$
6. $y = \sqrt{2} \cos x$
7. $y = \sin 2x$
8. $y = \frac{1}{2} \sin 3x$
9. $y = 1.2 \cos 4x$
10. $y = \frac{1}{2}\sqrt{3} \cos \left(\frac{4}{3} x\right)$

Sketch the graph of each of the following over the interval $0 \leq x \leq 2\pi$:

11. $f(x) = \frac{3}{5} \cos 2x$
12. $f(x) = \frac{3}{10} \sin 3x$
13. $f(x) = \frac{6}{5} \sin \frac{1}{2} x$
14. $f(x) = \frac{4}{5} \cos \frac{2}{3} x$

In problems 15 through 19, use a horizontal scale of 5 squares = 1 unit to draw the graph of each of the functions over one cycle.

EXAMPLE 5 $y = \sin \pi x$

SOLUTION   The period of the graph is $2\pi/\pi = 2$, so the zeros are at 0, 1, and 2. The maximum value of 1 occurs when $x = 0.5$, and the minimum value, -1, occurs when $x = 1.5$. The graph is shown in Figure 4–9.

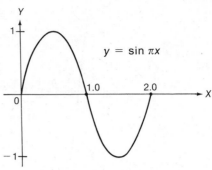

FIGURE 4–9

15. $y = \cos \pi x$
16. $y = \frac{4}{5} \sin \left(\frac{\pi}{2} x\right)$
17. $y = \frac{9}{10} \sin \left(\frac{2\pi}{3} x\right)$
18. $y = \frac{1}{2}\sqrt{3} \cos \left(\frac{5\pi}{6} x\right)$
19. $y = \sin x$ over the interval $0 \leq x \leq 3\pi/2$

20. Draw the graph of $y = \sin x$ again, using the same axes as in problem 19, only this time use a scale of $\pi/12$ radians = 1 square on the horizontal axis.

21. Draw the graph of $y = \pm\sqrt{9 - x^2}$ over the interval $-3 \leq x \leq 3$
    a. using a scale of 2 squares = 1 unit on both the horizontal and vertical axes.
    b. using a scale of 2 squares = 1 unit on the horizontal axis and 1 square = 1 unit on the vertical axis.

22. Draw the graph of $y = \pm\frac{2}{3}\sqrt{9 - x^2}$ over the interval $-3 \leq x \leq 3$
    a. using a scale of 2 squares = 1 unit on both the horizontal and vertical axes.
    b. using a scale of 2 squares = 1 unit on the horizontal axis and 3 squares = 1 unit on the vertical axis.

4.2 GRAPHS OF $y = A \sin B(x + C)$ AND $y = A \cos B(x + C)$

In the previous section we learned how to sketch the graphs of sine and cosine curves with different periods and amplitudes. In each case the starting point of the graph was $x = 0$. One of the applications of sine curves or sine waves is in electronics and electricity. It is possible to display a sine wave on a TV-like instrument called an oscilloscope by passing an alternating current through the set.

The alternation of the current, from peak to peak, causes the sine wave, and it is possible to carry out analyses of the current by means of the oscilloscope. There are dials on the oscilloscope which permit the operator to change the amplitude, the period (or number of cycles), and to shift the curve horizontally from left to right or vice versa. The graphs of $y = A \sin B(x + C)$ and $y = A \cos B(x + C)$ are called SINUSOIDAL CURVES.

EXAMPLE 1 Plot the graph of $y = \sin(x + \pi/6)$ over the interval $0 \le x \le 2\pi$.

SOLUTION Prepare a table of values, then plot the points, as in Figure 4–10.

x	0	$\dfrac{\pi}{12}$	$\dfrac{\pi}{6}$	$\dfrac{\pi}{4}$	$\dfrac{\pi}{3}$	$\dfrac{\pi}{2}$	$\dfrac{7\pi}{12}$	$\dfrac{2\pi}{3}$	$\dfrac{5\pi}{6}$
$x + \dfrac{\pi}{6}$	$\dfrac{\pi}{6}$	$\dfrac{\pi}{4}$	$\dfrac{\pi}{3}$	$\dfrac{5\pi}{12}$	$\dfrac{\pi}{2}$	$\dfrac{2\pi}{3}$	$\dfrac{3\pi}{4}$	$\dfrac{5\pi}{6}$	π
$\sin\left(x + \dfrac{\pi}{6}\right)$	0.50	0.71	0.87	0.97*	1.00	0.87	0.71	0.50	0

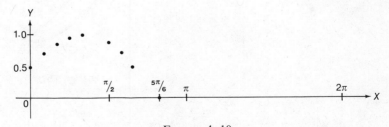

FIGURE 4–10

We were asked to plot the graph over $0 \le x \le 2\pi$ and we certainly could continue in this way, but it is extremely tedious. Let us recall some facts about the basic sine curve from previous sections and save some effort. First, the zeros of $y = \sin \alpha$ occur when $\alpha = k \cdot \pi$, or at $0, \pi, 2\pi, 3\pi$, and

* $(5\pi/12) = 75°$, $\sin 75° \approx 0.9659$

so on. This means that the zeros of $y = \sin(x + \pi/6)$ must occur when $x + \pi/6 = k \cdot \pi$, or when $x = k \cdot \pi - \pi/6$, which means at $0 - \pi/6 = -\pi/6$, at $\pi - \pi/6 = 5\pi/6$, and at $2\pi - \pi/6 = 11\pi/6$. The maximum values of $y = \sin \alpha$ occur when $\alpha = \pi/2 + k \cdot 2\pi$, so the maximum values of $y = \sin(x + \pi/6)$ occur when $x + \pi/6 = \pi/2 + k \cdot 2\pi$, or when $x = (\pi/2 - \pi/6) = k \cdot 2\pi = \pi/3 + k \cdot 2\pi$. Thus there will be a maximum value at $\pi/3$, another at $7\pi/3$, and so on. The minimum values of $y = \sin \alpha$ occur when $\alpha = 3\pi/2 + k \cdot 2\pi$, so the minimum values of $y = \sin(x + \pi/6)$ will occur when $x + \pi/6 = 3\pi/2 + k \cdot 2\pi$, or when $x = (3\pi/2 - \pi/6) + k \cdot 2\pi = 4\pi/3 + k \cdot 2\pi$. There will be a minimum value at $4\pi/3$, another at $10\pi/3$, and so on. The minimum value at $4\pi/3$ is found by replacing x with $4\pi/3$, so

$$\sin\left(\frac{4\pi}{3} + \frac{\pi}{6}\right) = \sin \frac{3\pi}{2} = -1$$

Similarly the maximum value, at $x = \pi/3$, is $\sin(\pi/3 + \pi/6) = \sin \pi/2 = 1$. The value of $\sin(x + \pi/6)$ when $x = 2\pi$ is $\sin(2\pi + \pi/6) = \sin \pi/6 = 0.5$. With this information, together with the table of values already determined, we obtain the graph shown in Figure 4–11. We have also superimposed the graph of $y = \sin x$ on that of $y = \sin(x + \pi/6)$, and it is evident that each point of $y = \sin x$ has been shifted $\pi/6$ units to the left.

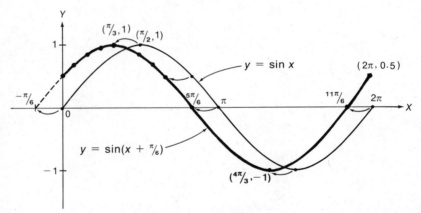

FIGURE 4–11

If we were to draw the graph of $y = \sin(x - \pi/6)$, the zeros would be where $x - \pi/6 = k \cdot \pi$, or at $x = \pi/6 + k \cdot \pi$, that is, at $\pi/6, 7\pi/6, 13\pi/6$, and so on. The maximums would occur when $x - \pi/6 = \pi/2 + k \cdot 2\pi$, or $x = 2\pi/3 + k \cdot 2\pi$. The minimums would occur when $x - \pi/6 = 3\pi/2 + k \cdot 2\pi$, or $x = 5\pi/3 + k \cdot 2\pi$. This graph, along with that of $y = \sin x$, is shown in Figure 4–12. Again it is evident that each point of $y = \sin x$ has been shifted $\pi/6$ units, this time to the right. In fact the graph of any equation of the form $y = \sin(x + C)$ or $y = \cos(x + C)$ is merely the basic curve shifted C units to the right if C is negative or C units to the left if C is positive.

4.2 GRAPHS OF $y = A \sin B(x + C)$ AND $y = A \cos B(x + C)$

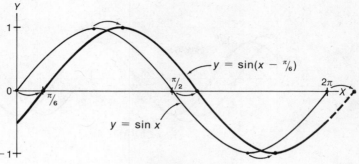

FIGURE 4–12

Next let us draw the graph of something more complicated, such as $y = 3/5 \cos 2(x - \pi/6)$. First we shall establish a few values for a plotting table, as in Table 4–1. It doesn't take much of this to see that a little bit of algebra applied

TABLE 4–1

x	0	$\dfrac{\pi}{6}$	$\dfrac{\pi}{3}$	$\dfrac{\pi}{2}$
$x - \dfrac{\pi}{6}$	$-\dfrac{\pi}{6}$	0	$\dfrac{\pi}{6}$	$\dfrac{\pi}{3}$
$2\left(x - \dfrac{\pi}{6}\right)$	$-\dfrac{\pi}{3}$	0	$\dfrac{\pi}{3}$	$\dfrac{2\pi}{3}$
$\cos 2\left(x - \dfrac{\pi}{6}\right)$	0.50	1.00	0.50	-0.50
$\dfrac{3}{5} \cos 2\left(x - \dfrac{\pi}{6}\right)$	0.30	0.60	0.30	-0.30

as in the previous example could save us a considerable amount of work. So if we consider that the zeros of $y = \cos \alpha$ occur when $\alpha = \pi/2 + k \cdot \pi$, the zeros of $y = 3/5 \cos 2(x - \pi/6)$ occur when $2(x - \pi/6) = \pi/2 + k \cdot \pi$, or when $x - \pi/6 = \pi/4 + k \cdot \pi/2$, which means when $x = 5\pi/12 + k \cdot \pi/2$. The maximum values of $y = \cos \alpha$ are reached when $\alpha = k \cdot 2\pi$, so those of $y = 3/5 \cos 2(x - \pi/6)$ happen when $2(x - \pi/6) = k \cdot 2\pi$, or $x - \pi/6 = k \cdot \pi$, or more simply when $x = \pi/6 + k \cdot \pi$. The maximum value when $x = \pi/6$ is $y = 3/5 \cos 2(\pi/6 - \pi/6) = 3/5 \cos (2 \cdot 0) = 3/5 \cos 0 = 3/5 \cdot (1) = 3/5$. The minimum values of $y = \cos \alpha$ occur when $\alpha = \pi + k \cdot 2\pi$, so for $y = 3/5 \cos 2(x - \pi/6)$ this means when $2(x - \pi/6) = \pi + k \cdot 2\pi$, or when $x - \pi/6 = \pi/2 + k \cdot \pi$, or $x = 2\pi/3 + k \cdot \pi$. The minimum value is $-3/5$. Using this information and the few points from the plotting table, we obtain the graph shown in Figure 4–13. This time we have superimposed the graph of $y = 3/5 \cos 2x$ on the same pair of axes, and it is evident that the graph of $y = 3/5 \cos 2(x - \pi/6)$ is exactly the same as that of $y = 3/5 \cos 2x$, except that each point is shifted $\pi/6$ units to the right.

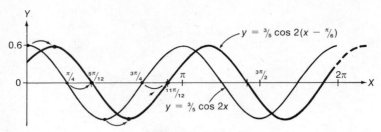

FIGURE 4–13

It can be shown that the graph of $y = A \sin B(x + C)$ or $y = A \cos B(x + C)$ is identical to that of $y = A \sin Bx$ or $y = A \cos Bx$, respectively, shifted C units, to the left if C is positive, to the right if C is negative. The number $|C|$ is called the PHASE SHIFT of the equation. Thus we have a means of sketching the graph of any equation of the form $y = A \sin B(x + C)$ or $y = A \cos B(x + C)$.

1. Determine the amplitude, $|A|$
2. Determine the period, $2\pi/|B|$
3. Sketch one cycle of $y = A \sin Bx$ or $y = A \cos Bx$
4. Shift each point $|C|$ units, left if C is positive, right if C is negative.

EXAMPLE 2 Sketch the graph of $y = \frac{1}{2}\sqrt{3} \sin 3(x + \pi/12)$ where $0 \le x \le 2\pi$.

SOLUTION The amplitude is $\frac{1}{2}\sqrt{3} \approx 0.87$. The period is $2\pi/3$. The phase shift is $\pi/12$, to the left since C is positive. Sketch one cycle of $y = \frac{1}{2}\sqrt{3} \sin 3x$, then shift each point $\pi/12$ units to the left. Note the relationship between maximum, zeros, and minimum and complete the graph, as in Figure 4–14. If you want to find where the graph begins and ends in the interval, you can simply substitute $x = 0$ to obtain $y =$

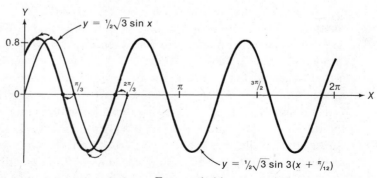

FIGURE 4–14

$\frac{1}{2}\sqrt{3} \sin 3(0 + \pi/12) = \frac{1}{2}\sqrt{3} \sin (\pi/4) = \frac{1}{2}\sqrt{3}(\frac{1}{2}\sqrt{2}) = \frac{1}{4}\sqrt{6} \approx \frac{1}{4}(2.449) \approx 0.61$, and substitute $x = 2\pi$ to find $y = \frac{1}{2}\sqrt{3} \sin 3(2\pi + \pi/12) = \frac{1}{2}\sqrt{3} \sin (3 \cdot 2\pi + \pi/4) = \frac{1}{2}\sqrt{3} \sin \pi/4 \approx 0.61$.

EXERCISES

1. Given $f(x) = 2 \cos 2(x - \pi/4)$, find the exact value of a. $f(0)$ b. $f(\pi/4)$ c. $f(\pi)$ d. $f(2\pi)$

2. Given $f(x) = \sqrt{3} \sin \frac{1}{2}(x + \pi/6)$, find the exact value of a. $f(\pi/6)$ b. $f(\pi/2)$ c. $f(4\pi/3)$ d. $f(2\pi)$

Sketch the graph of each of the following where $0 \leq x \leq 2\pi$:

3. $y = \sqrt{2} \sin (x + \pi/4)$
4. $y = 6/5 \sin (x - \pi/3)$
5. $y = 2 \cos (x - \pi/6)$
6. $y = \frac{1}{2} \cos (x + \pi/4)$
7. $y = \cos (x - \pi/2)$
8. $y = \sin (x + \pi/2)$
9. $y = 4/5 \sin 2(x - \pi/4)$
10. $y = 3/10 \cos 3(x - \pi/12)$
11. $y = \cos (2x - \pi/3)$. *Hint:* Change to the form $y = \cos B(x + C)$ first.

12. $y = \sin (3x + \pi/2)$

Sketch the graph of each of the following over one complete cycle. Use a horizontal scale of 5 squares = 1 unit.

13. $f(x) = \sin \pi(x - 0.2)$
14. $f(x) = \cos \frac{\pi}{3}(x + 0.1)$
15. $f(x) = 1.2 \cos \frac{\pi}{3}(x + 1.0)$
16. $f(x) = 0.8 \sin \pi(2x - 2)$

Prepare a plotting table at intervals of $\pi/6$ units and draw the graph of each of the following over the interval $0 \leq x \leq 2\pi$:

17. $f(x) = \sin x + \cos x$ *(Compare the result with the graph of problem 3.)*
18. $f(x) = \sin x + \sqrt{3} \cos x$ *(Compare the result with the graph of problem 5.)*

4.3 SKETCHING THE NONSINUSOIDAL TRIGONOMETRIC FUNCTIONS

Examination of the basic graphs of the remaining trigonometric functions and attention to some of the significant points, as indicated in Figure 4–15, will assist in the task of rapid sketching of $y = A \tan B(x + C)$, and so on.

Although none of these graphs has a maximum or minimum value, so that there is no amplitude as in the sine and cosine curves, the secant and cosecant have *relative* maximum and *relative* minimum values occurring at $(2k + 1)\pi$ and $k \cdot 2\pi$ for secant, at $(4k - 1) \cdot \pi/2$ and $(4k + 1) \cdot \pi/2$ for cosecant. The zeros of $y = A \tan x$ and $y = A \cot x$ at $k \cdot \pi$ and $(2k + 1) \cdot \pi/2$ are important, and both of these pass through the points $(\pi/4, A)$ and $(3\pi/4, -A)$. The period of $y = A \tan x$ and $y = A \cot x$ is π, whereas the period of $y = A \sec x$ and $y = A \csc x$

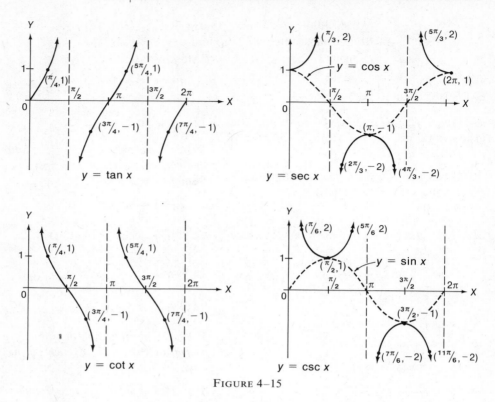

FIGURE 4-15

is 2π. Note the asymptotes of each curve, occurring at $(2k + 1) \cdot \pi/2$ for tangent and secant, at $k \cdot \pi$ for cotangent and cosecant. (If the distance between a curve and a fixed straight line approaches zero, the straight line is an ASYMPTOTE of the curve.)

EXAMPLE 1   Sketch the graph of $y = 3/5 \tan x$.

SOLUTION   The zeros and asymptotes are the same as $y = \tan x$, so locate these as in Figure 4-16. When $x = \pi/4$, $y = 3/5 \tan \pi/4 = (3/5) \cdot 1 = 3/5$, and at $3\pi/4$, $y = 3/5 \tan 3\pi/4 = (3/5)(-1) = -3/5$. Plot these points and sketch the graph.

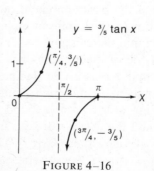

FIGURE 4-16

EXAMPLE 2   Sketch the graph of $y = \frac{1}{2} \csc x$.

SOLUTION   Asymptotes remain the same as for $y = \csc x$. The relative minimum occurs at $\pi/2$, where $y = \frac{1}{2} \csc \pi/2 = \frac{1}{2}(1) = \frac{1}{2}$, and the relative maximum is at $3\pi/2$, since $y = \frac{1}{2} \csc 3\pi/2 = \frac{1}{2}(-1) = -\frac{1}{2}$. If more accuracy is desired, use $\pi/6, 5\pi/6, 7\pi/6$, and $11\pi/6$ to get $y = \frac{1}{2} \csc \pi/6 = \frac{1}{2}(2) = 1$, $y = \frac{1}{2} \csc 5\pi/6 = \frac{1}{2}(2) = 1$, $y = \frac{1}{2} \csc 7\pi/6 = \frac{1}{2}(-2) = -1$, and $y = \frac{1}{2} \csc 11\pi/6 = \frac{1}{2}(-2) = -1$. Plot these points, draw the asymptotes, then sketch the graph as in Figure 4–17.

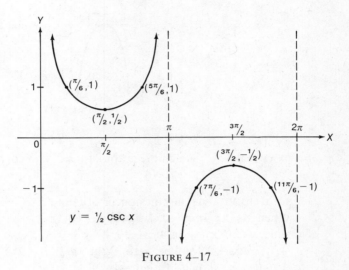

FIGURE 4–17

The effect of $B \neq 1$ in graphing of $y = A \sec Bx$ and $y = A \csc Bx$ is to change the period from 2π to $2\pi/|B|$. The location of relative maximum and minimum values and asymptotes is also determined by division by $|B|$.

EXAMPLE 3   Sketch the graph of $y = \sec 2x$.

SOLUTION   The asymptotes of $y = \sec x$, which are located at $x = \pi/2$ and $x = 3\pi/2$, now become $x = \pi/4$ and $x = 3\pi/4$, while the relative minimums are now at 0 and π, and the relative maximum is at $\pi/2$. To obtain a more accurate graph we can locate $(\pi/6, 2), (\pi/3, -2), (2\pi/3, -2)$, and $(5\pi/6, 2)$. Sketch as in Figure 4–18 on the next page.

The effect of $B \neq 1$ in graphing $y = A \tan Bx$ and $y = A \cot Bx$ is to change the period from π to $\pi/|B|$. Locations of asymptotes and points where $y = \pm A$ are also changed by division by $|B|$. Remember that $A \tan \pi/4 = A \cdot 1 = A$.

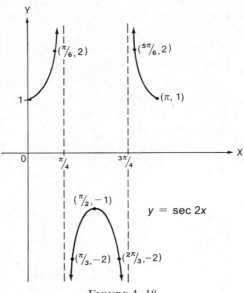

FIGURE 4–18

EXAMPLE 4   Sketch the graph of $y = \frac{1}{2}\sqrt{3} \cot \pi x$.

SOLUTION   Asymptotes occur at 0 and $\pi/\pi = 1$ rather than at 0 and π. When $x = (\pi/4)/\pi = 1/4$, $y = \frac{1}{2}\sqrt{3} \cot(\pi \cdot \frac{1}{4}) = \frac{1}{2}\sqrt{3}(1) = \frac{1}{2}\sqrt{3} \approx 0.87$, and when $x = (3\pi/4)/\pi = 3/4$, $y = \frac{1}{2}\sqrt{3} \cot(\pi \cdot \frac{3}{4}) = \frac{1}{2}\sqrt{3}(-1) = -\frac{1}{2}\sqrt{3} \approx -0.87$. The zero is at $(\pi/2)/\pi = 1/2$. Two cycles are easily sketched, as shown in Figure 4–19.

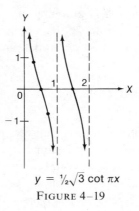

$y = \frac{1}{2}\sqrt{3} \cot \pi x$
FIGURE 4–19

Graphs of all four nonsinusoidal functions shift right or left in the same way as the sine and cosine curves.

4.3 SKETCHING THE NONSINUSOIDAL TRIGONOMETRIC FUNCTIONS

EXAMPLE 5 Sketch the graph of $y = 0.3 \tan (x + \pi/6)$ over the interval $0 \le x \le 2\pi$.

SOLUTION First sketch the graph of $y = 0.3 \tan x$. Each point is shifted $\pi/6$ units to the left. Since $x + \pi/6 = \pi/4$ means that $x = \pi/4 - \pi/6 = \pi/12$, you can check by finding $f(\pi/12) = 0.3 \tan (\pi/12 + \pi/6) = 0.3 \tan \pi/4 = 0.3$. The results are shown in Figure 4–20.

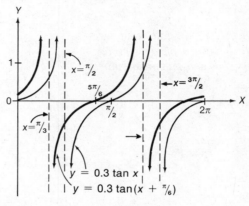

FIGURE 4–20

EXAMPLE 6 Sketch the graph of $y = \dfrac{4}{5} \sec \dfrac{\pi}{2}(x - 0.3)$ over the interval $0 \le x \le 5$.

SOLUTION First sketch the graph of $y = \dfrac{4}{5} \sec \dfrac{\pi}{2} x$. The period is $2\pi/(\pi/2) = 4$. Then shift each point 0.3 units to the right. The graph is shown in Figure 4–21.

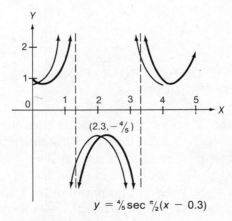

FIGURE 4–21

EXERCISES

1. Given $f(x) = 2 \sec 3x$, find the exact value of
   a. $f(0)$
   b. $f(\pi/12)$
   c. $f(\pi/6)$
   d. $f(\pi/4)$
   e. $f(\pi/3)$

2. Given $g(x) = \sqrt{3} \cot \dfrac{1}{2} x$, find the exact value of
   a. $g(0)$
   b. $g(\pi/3)$
   c. $g(\pi/2)$
   d. $g(\pi)$
   e. $g(5\pi/3)$

Sketch the graph of each of the following over one cycle. Indicate locations of asymptotes, zeros, relative maximum and minimum values as appropriate.

3. $y = \dfrac{4}{5} \sec x$
4. $y = \dfrac{2}{5} \cot x$

5. $y = \csc 2x$
6. $y = \tan \dfrac{3}{2} x$

7. $y = \dfrac{1}{2} \sqrt{2} \tan \dfrac{\pi}{2} x$

8. $y = \dfrac{9}{10} \cot \dfrac{2\pi}{3} x$

9. $y = \cot \left(x - \dfrac{\pi}{12} \right)$

10. $y = \sec 2 \left(x + \dfrac{\pi}{6} \right)$

Sketch the graph of each of the following over the interval $0 \leq x \leq 2\pi$. Indicate locations of asymptotes, zeros, relative maximum and minimum values as appropriate.

11. $y = \dfrac{7}{10} \sec 3 \left(x + \dfrac{\pi}{6} \right)$

12. $y = \dfrac{1}{2} \csc 3 \left(x - \dfrac{\pi}{4} \right)$

Sketch the graph of each of the following over the interval $-\pi \leq x \leq \pi$. Indicate locations of asymptotes, zeros, relative maximum and minimum values as appropriate.

13. $f(x) = -\sin x$
14. $f(x) = -\cos x$

15. $f(x) = \dfrac{1}{2} \cos 2x$
16. $f(x) = \dfrac{3}{5} \sin 3x$

17. $f(x) = \cot \dfrac{3}{2} x$
18. $f(x) = \csc \dfrac{4}{3} x$

19. $f(x) = \dfrac{4}{5} \sin \left(x - \dfrac{\pi}{4} \right)$

20. $f(x) = \sqrt{2} \tan \left(x - \dfrac{\pi}{12} \right)$

4.4 GRAPHS OF RELATIONS, FUNCTIONS, AND THEIR INVERSES

In algebra a RELATION is defined as a set of ordered pairs (x, y). If $A = \{(x, y)\}$, then $A^{-1} = \{(y, x)\}$ is the INVERSE of the relation A. The relation may be a very simple set of ordered pairs, such as $A = \{(-3, -1), (-1, 0), (0, 3), (2, 4)\}$. Then $A^{-1} = \{(-1, -3), (0, -1), (3, 0), (4, 2)\}$. The graphs of A and A^{-1} are shown in Figure 4–22. We have also drawn in the graph of the line $y = x$. Note that the points of A and A^{-1} are symmetric with respect to this line, a fact which is true of any relation and its inverse. Proof of this is left as an exercise at the end of the section.

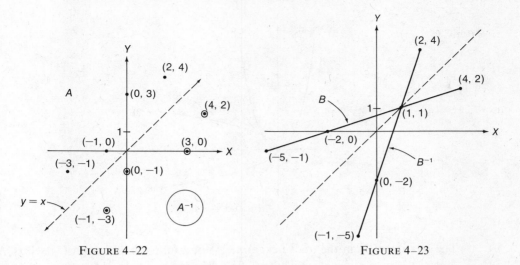

FIGURE 4–22   FIGURE 4–23

Another relation, $B = \{(x, y) \mid y = \frac{1}{3}(x + 2) \text{ and } -5 \leq x \leq 4\}$, is graphed in Figure 4–23. The points $(-5, -1)$, $(-2, 0)$, $(1, 1)$, and $(4, 2)$ are part of B, so its inverse, B^{-1}, contains $(-1, -5)$, $(0, -2)$, $(1, 1)$, and $(2, 4)$. However B^{-1} should include all points between $(-1, -5)$ and $(2, 4)$, and we would expect the graph to be a line segment, as is the case. By various means we could find an equation to represent B^{-1}, but the easiest way is simply to interchange x and y in the equation defining B to obtain $x = \frac{1}{3}(y + 2)$. If we want to write the equation for y in terms of x, we can solve this equation for y and find $y = 3x - 2$.

Another fact to be noted here is that we arbitrarily limited the domain of B, so that $\mathscr{D}_B = \{x \mid -5 \leq x \leq 4\}$. The domain of a relation is the set of first elements in the relation. This necessarily limited the range of B to the set $\mathscr{R}_B = \{y \mid -1 \leq y \leq 2\}$. The range of a relation is the set of second elements in the relation. From the graph of B^{-1} we see that $\mathscr{D}_{B^{-1}} = \{x \mid -1 \leq x \leq 2\}$ and $\mathscr{R}_{B^{-1}} = \{y \mid -5 \leq y \leq 4\}$. It is evident, then, that the range of the inverse is the domain of the original relation and that the domain of the inverse is the range of the original relation. More simply, domains and ranges are interchanged.

EXAMPLE 1   Draw the graph of $C = \{(x, y) \mid y = x^2, -3 \leq x \leq 3\}$, then   *a.* find the range of C *b.* write an equation for C^{-1} *c.* determine the domain and range of C^{-1} *d.* draw the graph of C^{-1} on the same pair of axes with C.

SOLUTION   See Figure 4–24 for the graph.   *a.* From the graph we see that $\mathscr{R}_C = \{y \mid 0 \leq y \leq 9\}$.   *b.* Interchanging x and y in the equation defining C, we obtain $x = y^2$.   *c.* Interchanging domains and ranges, $\mathscr{D}_{C^{-1}} = \{x \mid 0 \leq x \leq 9\}$ and $\mathscr{R}_{C^{-1}} = \{y \mid -3 \leq y \leq 3\}$.   *d.* The graph is shown in Figure 4–24.

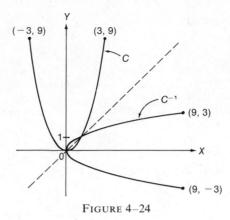

FIGURE 4–24

The relation C^{-1} in the above example is not a function, although C itself is. A function is a relation in which each first element has only one second element. If the equation for C^{-1} is solved for y, then $x = y^2$ is equivalent to $y = \pm\sqrt{x}$, so we see that for all $x > 0$, y has two distinct values.

Sometimes it is desirable to restrict the domain of a given function in order that its inverse will also be a function. For example, we can define $D = \{(x, y) \mid y = x^2$ and $x \geq 0\}$, so that $D^{-1} = \{(x, y) \mid y = \sqrt{x}\}$. The graphs of D and D^{-1} are shown in Figure 4–25. Here we simply chose the equation $y = \sqrt{x}$ rather than $y = -\sqrt{x}$ when given the choice $y = \pm\sqrt{x}$ after solving $x = y^2$ for y. We could define a function $E = \{(x, y) \mid y = x^2, x \leq 0\}$, however, which has $E^{-1} = \{(x, y) \mid y = -\sqrt{x}\}$, and with graphs as shown in Figure 4–26.

EXAMPLE 2   Draw the graph of the relation $F = \{(x, y) \mid y = \frac{1}{2}\sqrt{16 - x^2}\}$ and its inverse, F^{-1}, then restrict the domain of F in such a way that F^{-1} will be a function.

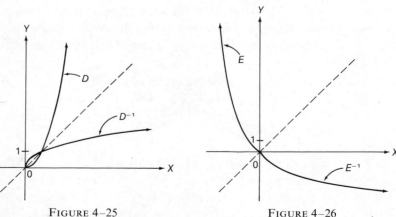

FIGURE 4–25   FIGURE 4–26

4.4 GRAPHS OF RELATIONS, FUNCTIONS, AND THEIR INVERSES

SOLUTION Draw the graphs of F and F^{-1} as shown in Figure 4–27. Note that $\mathscr{D}_F = \{x \mid -4 \leq x \leq 4\}$ and that F is a semiellipse. There are many choices for $\mathscr{D}_F$ which will make F^{-1} a function. Two choices would be $\mathscr{D}_F = \{x \mid 0 \leq x \leq 4\}$ or $\mathscr{D}_F = \{x \mid -4 \leq x \leq 0\}$. The graphs of F and F^{-1} for each of these domains are shown in Figure 4–28a and b.

We can write equations for F^{-1} in the following manner: interchange x and y as in previous examples to obtain $x = \frac{1}{2}\sqrt{16 - y^2}$. This would naturally restrict the domain of F^{-1} to $\{x \mid x \geq 0\}$, since the radical is always non-negative. The range would be $\{y \mid -4 \leq y \leq 4\}$. In order to solve for y in terms of x we must square both sides of the equation to obtain $x^2 = \frac{1}{4}(16 - y^2)$. Simplifying, we get $4x^2 + y^2 = 16$, the equation of an ellipse. Solving this for y, we find $y^2 = 16 - 4x^2$, or $y = \pm 2\sqrt{4 - x^2}$. Thus if we have $F = \{(x, y) \mid y = \frac{1}{2}\sqrt{16 - x^2} \text{ and } 0 \leq x \leq 4\}$, then $F^{-1} = \{(x, y) \mid y = 2\sqrt{4 - x^2}\}$, whereas if $F = \{(x, y) \mid y = \frac{1}{2}\sqrt{16 - x^2} \text{ and } -4 \leq x \leq 0\}$, $F^{-1} = \{(x, y) \mid y = -2\sqrt{4 - x^2}\}$.

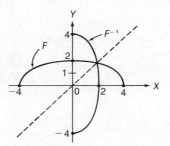

FIGURE 4–27

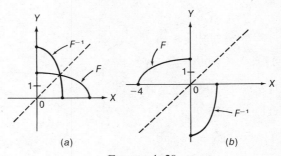

FIGURE 4–28

EXAMPLE 3 Draw the graph of $G = \{(x, y) \mid y = 2^x \text{ and } -3 \leq x \leq 3\}$, then draw the graph of G^{-1}.

SOLUTION Refer to Figure 4–29. The graphing is simple enough to do. Here, the equation of the inverse, $x = 2^y$, cannot be solved for y in terms of x by algebraic methods. If we want such an equation we must choose a name for the inverse—in this case $y = \log_2 x$. In the following section we shall graph the trigonometric functions and their inverses.

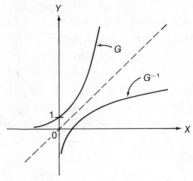

FIGURE 4–29

EXERCISES

Draw the graph of each of the following relations and their inverses, using a separate pair of axes for each relation-inverse pair. In each case tell whether or not the relation or inverse is a function, and give the domain and range of both.

1. $H = \{(-5, 0), (-3, -1), (-1, 0), (1, 3), (3, 4)\}$

2. $I = \{(-2, -3), (-1, -2), (-1, 0), (0, 2), (1, 3), (4, 1)\}$

3. $J = \{(x, y) \mid y = 3x \text{ and } x \geq 0\}$

4. $K = \{(x, y) \mid y = x^3 \text{ and } -2 \leq x \leq 2\}$

5. $L = \{(x, y) \mid y = x^{1/3} \text{ and } -8 \leq x \leq 8\}$

6. $M = \{(x, y) \mid y = \sqrt{25 - x^2}\}$

7. $N = \{(x, y) \mid x^2 + y^2 = 36 \text{ and } 0 \leq x \leq 6\}$

8. $P = \{(x, y) \mid y = (x + 2)^2 \text{ and } -4 \leq x \leq 0\}$

9. $Q = \{(x, y) \mid 9x^2 + y^2 = 36 \text{ and } -2 \leq x \leq 0\}$

10. $R = \{(x, y) \mid y = 12/x \text{ and } -12 \leq x \leq -1\}$

11. Refer to the graph of N in problem 7. Write an equation for N^{-1} with y in terms of x which will make N^{-1} a function.

12. Refer to the graph of P in problem 8. Write an equation for P^{-1} with y in terms of x which will make P^{-1} a function.

13. Draw the graph of the inverse of each of the relations shown in Figure 4–30.

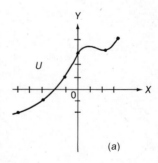

(a)

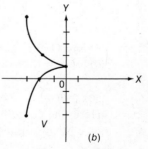

(b)

FIGURE 4–30

14. If $P(a, b)$ is an element of S, then $Q(b, a)$ is an element of S^{-1}. Prove that (a, b) and (b, a) are symmetric with respect to the line $y = x$, if $a \neq b$. Hint: If P and Q are symmetric with respect to a line, then the line is the perpendicular bisector of segment PQ. Show that the slopes of PQ and $y = x$ are negative reciprocals, and that if $R(x_1, x_1)$ is a point on $y = x$, the distance RP equals the distance RQ.

4.5 THE INVERSE TRIGONOMETRIC FUNCTIONS

We should already be able to draw the graphs of the inverses of the trigonometric functions. All we need do is consider the graph of the basic function, and then find the inverse points, as was done in Figure 4–31. We could write the equation of the inverse, $x = \sin y$, and find values for x and y, but the set of points we plotted for $y = \sin x$ was $\{(-\pi, 0), (-\pi/2, -1), (0, 0), (\pi/2, 1), (\pi, 0), (3\pi/2, 1), (2\pi, 0), (5\pi/2, 1), (3\pi, 0)\}$, so for its inverse we plotted the set $\{(0, -\pi), (-1, -\pi/2), (0, 0), (1, \pi/2), (0, \pi), (1, 3\pi/2), (0, 2\pi), (1, 5\pi/2), (0, 3\pi)\}$. From this it is obvious that the inverse of the function $S = \{(x, y) \mid y = \sin x\}$ is not a function. The domain and range of S, we might observe, are the set of all real numbers and $\{y \mid -1 \leq y \leq 1\}$, respectively, while the domain and range of S^{-1} are $\{x \mid -1 \leq x \leq 1\}$ and the set of all real numbers.

In order to have an inverse trigonometric function, then, we must restrict the domain of the basic function so that the inverse will be a function. This can be accomplished arbitrarily in many ways, but the domains we use here are fairly universally accepted. For example, we define $y = \mathrm{Sin}\, x$, with domain $\{x \mid -\pi/2 \leq x \leq \pi/2\}$ and range $\{y \mid -1 \leq y \leq 1\}$, so that $x = \mathrm{Sin}\, y$ will have domain $\{x \mid -1 \leq x \leq 1\}$ and range $\{y \mid -\pi/2 \leq y \leq \pi/2\}$. Their graphs are shown in Figure 4–32. If we wish to change $x = \mathrm{Sin}\, y$ to a function of y in terms of x, we use the notation $y = \mathrm{Sin}^{-1} x$ or $y = \mathrm{Arc\,sin}\, x$. We prefer the latter

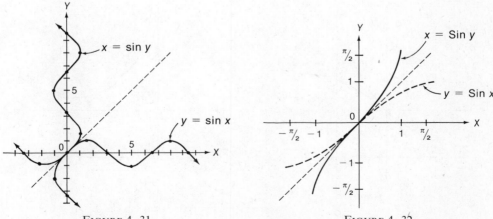

FIGURE 4–31   FIGURE 4–32

because the former might tend to be confused with the -1 power of Sin x or the reciprocal of Sin x.

The remaining five restricted functions with their domains and ranges, together with their inverses and their domains and ranges, are listed in Table 4–2.

TABLE 4–2

Function	Domain	Range	Inverse	Domain	Range
$y = \text{Cos } x$	$0 \leq x \leq \pi$	$-1 \leq y \leq 1$	$y = \text{Arc cos } x$	$-1 \leq x \leq 1$	$0 \leq y \leq \pi$
$y = \text{Tan } x$	$-\pi/2 < x < \pi/2$	$-\infty < y < \infty$	$y = \text{Arc tan } x$	$-\infty < x < \infty$	$-\pi/2 < y < \pi/2$
$y = \text{Cot } x$	$0 < x < \pi$	$-\infty < y < \infty$	$y = \text{Arc cot } x$	$-\infty < x < \infty$	$0 < y < \pi$
$y = \text{Sec } x$	$\begin{cases} -\pi \leq x < -\pi/2 \\ 0 \leq x < \pi/2 \end{cases}$	$\begin{cases} -\infty < y \leq -1 \\ 1 \leq y < \infty \end{cases}$	$y = \text{Arc sec } x$	$\begin{cases} -\infty < x \leq -1 \\ 1 \leq x < \infty \end{cases}$	$\begin{cases} -\pi \leq y < -\pi/2 \\ 0 \leq y < \pi/2 \end{cases}$
$y = \text{Csc } x$	$\begin{cases} -\pi < x \leq -\pi/2 \\ 0 < x \leq \pi/2 \end{cases}$	$\begin{cases} -\infty < y \leq -1 \\ 1 \leq y < \infty \end{cases}$	$y = \text{Arc csc } x$	$\begin{cases} -\infty < x \leq -1 \\ 1 \leq x < \infty \end{cases}$	$\begin{cases} -\pi < y \leq -\pi/2 \\ 0 < y \leq \pi/2 \end{cases}$

The graphs of $y = \text{Tan } x$ and $y = \text{Arc tan } x$, and of $y = \text{Csc } x$ and $y = \text{Arc csc } x$ are shown in Figure 4–33. The reader will be asked to draw the graphs of the remaining trigonometric functions and their inverses in the exercises at the end of the section. This may seem quite complicated, but the main thing to remember is the range of each of the inverse functions, that is, that

$$-\pi/2 \leq \text{Arc sin } x \leq \pi/2$$

$$-\pi/2 < \text{Arc tan } x < \pi/2$$

$$0 \leq \text{Arc cos } x \leq \pi$$

$$0 < \text{Arc cot } x < \pi$$

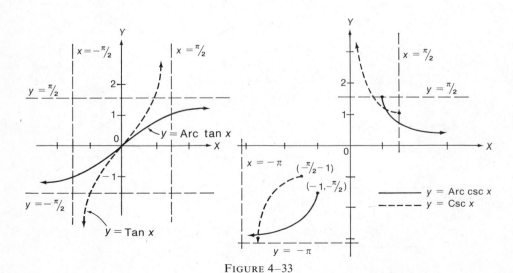

FIGURE 4–33

4.5 THE INVERSE TRIGONOMETRIC FUNCTIONS

The inverse secant and cosecant are not listed because you probably will not need to know them now.

We should note that the relation which is the inverse of $y = \sin x$, with equation $x = \sin y$, is sometimes written as $y = \text{arc sin } x$. There are also some books which use $y = \text{arc sin } x$ to refer to the inverse trigonometric function with range $-\pi/2 \leq \text{arc sin } x \leq \pi/2$, so this is simply a warning that elsewhere you may have to get the meaning from either definition or context.

Some of the exercises which follow are for the purpose of familiarizing you with the range of values of the inverse functions, others are useful in later mathematics courses, and some represent practical applications.

EXAMPLE 1   Find the value of each of the following:
a. $y = \text{Arc sin } (\frac{1}{2}\sqrt{3})$
b. $y = \text{Arc tan } (-1)$
c. $y = \text{Arc cos } (-\frac{1}{2})$
d. $\text{Arc cot } (0.3212)$

SOLUTION   a. Recall that $y = \text{Arc sin } x$ is equivalent to $x = \text{Sin } y$, with y restricted to the interval $-\pi/2 \leq y \leq \pi/2$. Thus if $y = \text{Arc sin } (\frac{1}{2}\sqrt{3})$, this is equivalent to the equation $\frac{1}{2}\sqrt{3} = \text{Sin } y$. When we solved equations of the type $\frac{1}{2}\sqrt{3} = \sin y$ in Section 3.5, y was an angle whose sine was $\frac{1}{2}\sqrt{3}$, and that might have been $\frac{\pi}{3} + k \cdot 2\pi$ or $\frac{2\pi}{3} + k \cdot 2\pi$. Now $y = \text{Arc sin } x$ is single-valued, however, so $y = \pi/3$, the *only* value within the range of the inverse sine function $(-\pi/2 \leq y \leq \pi/2)$.   b. $y = \text{Arc tan } (-1)$ can be thought of as the angle within the range $-\pi/2$ through $\pi/2$ whose tangent is -1. Thus, since $\tan \pi/4 = 1$ and $\tan (-\pi/4) = -1$, $y = \text{Arc tan } (-1) = -\pi/4$.   c. $y = \text{Arc cos } (-1/2)$ is the angle within the range 0 through π whose cosine is $-1/2$. Since $\cos \pi/3 = 1/2$ and $\cos 2\pi/3 = -1/2$, $y = \text{Arc cos } (-1/2) = 2\pi/3$.   d. We probably don't know the angle whose cotangent is 0.3212, so we look it up in Table II in Appendix D. (We generally prefer to have the answer in radians.) Thus $y = \text{Arc cot } (0.3212) = 1.26$.

EXAMPLE 2   Find the exact value of each of the following:
a. $\sin [\text{Arc cos } (-1/2)]$
b. $\cos [\text{Arc sin } (-1/2)]$
c. $\tan [\text{Arc sin } (\sqrt{5}/3)]$

SOLUTION   a. We determined in Example 1 that $\text{Arc cos } (-1/2) = 2\pi/3$. Thus $\sin [\text{Arc cos } (-1/2)] = \sin 2\pi/3 = (1/2)\sqrt{3}$.   b. The value of $\text{Arc sin } (-1/2) = -\pi/6$, so $\cos [\text{Arc sin } (-1/2)] = \cos (-\pi/6) = (1/2)\sqrt{3}$.   c. We could look up the angle whose sine is $\sqrt{5}/3$, or approximately so, in Table II in Appendix D. To find the exact value, however, we can let $\alpha = \text{Arc sin } \sqrt{5}/3$, as in Figure 4–34. Then if a point P is located on the terminal side of α such that $y = \sqrt{5}$ and $r = 3$, we find that $x^2 = 3^2 - (\sqrt{5})^2 = 9 - 5 = 4$, so that $x = 2$. Thus $\tan [\text{Arc sin } (\sqrt{5}/3)] = \tan \alpha = \sqrt{5}/2$. Another method would be to use the identity $\cos^2 \alpha = 1 - \sin^2 \alpha$

so that $\cos^2 \alpha = 1 - (\sqrt{5}/3)^2 = 1 - 5/9 = 4/9$, and $\cos \alpha = 2/3$. Then $\tan \alpha = \sin \alpha / \cos \alpha = (\sqrt{5}/3)/(2/3) = \sqrt{5}/2$.

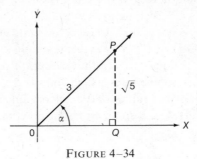

FIGURE 4–34

EXAMPLE 3   Find the volume of oil in a cylindrical tank with diameter 5 feet and 12 feet in length, lying on its side as in Figure 4–35, if the depth of the oil is 1 foot.

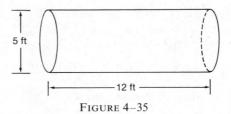

FIGURE 4–35

SOLUTION   The volume is found by multiplying the area of the cross section at the end by the length of the tank, so we are interested in the area of the circular segment ABC as indicated in Figure 4–36. In order to find this we must subtract the area of triangle AOC from the area of the circular sector which has angle AOC as its central angle. The area S of the sector can be found by means of the formula $S = \frac{1}{2} r^2 \theta$, which was developed earlier.* The area T of the triangle can be determined by finding one-half the product of OD and AC, where $OD = OB - BD$ and $CD = \sqrt{(OC)^2 - (OD)^2} = \frac{1}{2} AC$.

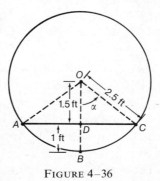

FIGURE 4–36

* Exercise 9, Chapter One Review Exercises, page 33.

4.5 THE INVERSE TRIGONOMETRIC FUNCTIONS

Note that $\theta = 2\alpha$, and that $\alpha = \text{Arc cos } (OD/r)$. Thus

$$V = 12(S - T)$$
$$= 12\left[\frac{1}{2}(2.5)^2 \cdot 2 \text{ Arc cos}\left(\frac{1.5}{2.5}\right) - (1.5)\sqrt{(2.5)^2 - (1.5)^2}\right]$$
$$= 12[(6.25) \text{ Arc cos } 0.6 - 1.50\sqrt{4.00}]$$
$$= 12[(6.25)(0.93) - 3.00] = 12(5.8125 - 3.00)$$
$$= 12(2.8125) = 33.75 \text{ cubic feet}$$

EXERCISES

Find the value of each of the following in radian measure:

1. Arc cos $(\frac{1}{2}\sqrt{3})$
2. Arc tan $\sqrt{3}$
3. Arc cot $\sqrt{3}$
4. Arc csc 2
5. Arc sin $(-\frac{1}{2}\sqrt{2})$
6. Arc cos $(-\frac{1}{2}\sqrt{2})$
7. Arc cos $(-\frac{1}{2}\sqrt{3})$
8. Arc sin $(-\frac{1}{2}\sqrt{3})$
9. Arc sin (-1)
10. Arc cos (-1)
11. Arc tan (0.6552)
12. Arc cos (0.9689)
13. Arc sin (-0.7707)
14. Arc tan (-0.5080)

Find the exact value of each of the following:

15. tan (Arc cos $\frac{1}{2}\sqrt{3}$)
16. sin (Arc tan $\sqrt{3}$)
17. tan (Arc cot $\sqrt{3}$)
18. sec (Arc cos $\frac{1}{2}\sqrt{2}$)
19. cos [Arc sin $(-\frac{1}{2}\sqrt{3})$]
20. sin [Arc tan (-1)]
21. cot [Arc cos $(-\frac{1}{2}\sqrt{3})$]
22. tan [Arc sin $(-1/2)$]
23. sin (Arc tan 3/4)
24. cos (Arc sin 12/13)
25. tan (Arc cos $\sqrt{11}/6$)
26. cos [Arc tan $(-\sqrt{13}/6)$]
27. tan (Arc tan 5/9)
28. cos [Arc cos $(-1/3)$]
29. sin [Arc cos $(-2/3)$]
30. cot [Arc sin $(-3/4)$]

Verify the following problems, as in Example 4.

EXAMPLE 4 Arc tan $1 +$ Arc tan $\dfrac{1}{2} =$ Arc tan 3

SOLUTION Let $\alpha = $ Arc tan 1, so tan $\alpha = 1$, and $\beta = $ Arc tan $\dfrac{1}{2}$, so tan $\beta = \dfrac{1}{2}$. Then

$$\tan(\alpha + \beta) = \frac{\tan\alpha + \tan\beta}{1 - \tan\alpha \cdot \tan\beta}$$
$$= \frac{1 + \frac{1}{2}}{1 - \frac{1}{2}} = \frac{3/2}{1/2} = 3$$

Thus

$$\text{Arc tan } 1 + \text{Arc tan } \frac{1}{2} = \text{Arc tan } 3$$

31. Arc tan $4 +$ Arc tan $(5/3) = 3\pi/4$
32. Arc tan $(4/3) -$ Arc tan $(1/7) = \pi/4$
33. Arc cos $(1/2) + 2$ Arc sin $(1/2) = 2\pi/3$

34. Arc cot 3 + Arc csc $\sqrt{5} = \pi/4$

35. Show that if $d = OC$, $\ell = AB$, and $r = OA$ in Figure 4–37, then $\theta = 2 \text{ Arc cos}(d/r) = 2 \text{ Arc tan}(\ell/2d) = 2 \text{ Arc sin}(\ell/2r)$

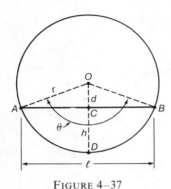

FIGURE 4–37

36. Show that if $h = CD$ and $r = OA$ as in Figure 4–37, the area S of segment ABD of the circle can be found by means of the formula

$$S = r^2 \text{ Arc cos}\left(\frac{r-h}{r}\right) - (r-h)\sqrt{2rh - h^2}$$

37. Referring to Figure 4–37, find θ in radians (nearest hundredth), if   a. $d = 10$ inches, $r = 12$ inches   b. $\ell = 2$ cm, $d = 10$ cm   c. $\ell = 5.20$ m, $r = 5.00$ m

38. One gallon of water fills approximately 231 cubic inches. Find the number of gallons of water in a cylindrical barrel lying on its side if it is 12 feet in length, the diameter of a circular end is 5 feet, and the depth of the water is   a. 6 inches   b. 24 inches   c. 30 inches   d. 36 inches

39. Draw the graphs of $y = \text{Cot } x$ and $y = \text{Arc cot } x$ on the same pair of axes.

40. Draw the graphs of $y = \text{Sec } x$ and $y = \text{Arc sec } x$ on the same pair of axes.

41. Draw the graphs of $y = \text{Cos } x$ and $y = \text{Arc cos } x$ on the same pair of axes.

4.6 ADDITION OF ORDINATES

Sometimes we need to draw the graphs of curves which are somewhat more complicated than those we have considered thus far. There are many natural phenomena such as sound, electrical waves, the orbiting of satellites, and a host of others, which can be analyzed by means of these simple curves and also by a combination of them. Joseph Fourier, a French mathematician who lived about 200 years ago, proved that *any* curve could be duplicated by addition of a sufficient number of periodic curves.

Graphing by means of plotting tables is rather tedious, so we can hasten matters by the methods of ADDITION OF ORDINATES, which will be demonstrated by means of several examples. First let us consider the graph of $y = 1 + \sin x$. This can be considered as the graph of $f(x) = g(x) + h(x)$, where $g(x) = 1$ and $h(x) = \sin x$. The graphs of $g(x)$ and $h(x)$ are drawn, then $f(x)$ is determined by plotting $f(a)$ from $g(a)$ and $h(a)$, as shown in Figure 4–38. Here we can see that $g(\pi/2) = 1$, $h(\pi/2) = 1$, and $f(\pi/2)$ will be $g(\pi/2) + h(\pi/2) = 1 + 1 = 2$, and so on, so that the maximum and minimum points are readily determined. The graph is simply that

4.6 ADDITION OF ORDINATES

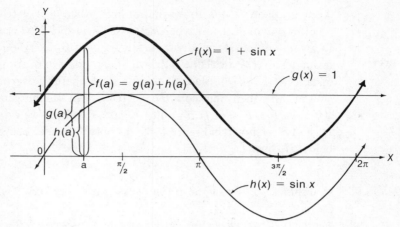

FIGURE 4–38

of $h(x) = \sin x$ moved up 1 unit everywhere. Things do become more complicated, however.

EXAMPLE 1   Draw the graph of $y = x + \sin x$ by addition of ordinates.

SOLUTION   This time let $g(x) = x$ and $h(x) = \sin x$. The graph in Figure 4–38 was drawn with the different horizontal and vertical scales recommended in Section 4.1, but in order to have $g(x) = x$ appear with the expected slope, a decimal scale is used in Figure 4–39. Observe here that when $h(x) >$

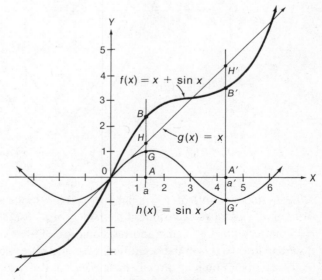

FIGURE 4–39

0, the graph of $f(x)$ is above that of $g(x)$, when $h(x) < 0$, the graph of $f(x)$ is below that of $g(x)$, and when $h(x) = 0$ the graphs of $f(x)$ and $g(x)$ intersect. Drawing compasses or dividers can be used to carry out the addition of ordinates. If a vertical line is drawn at any point a, then open the points the distance $AG = h(a)$, place one point at H where the vertical line intersects $g(x)$ at $g(a)$, then mark off $HB = AG = h(a)$ on the vertical line, so that $AB = AH + BH$, or $f(a) = g(a) + h(a)$. Similarly we can locate a', A', G', H', and B' as shown in Figure 4–39, the only difference being that B' is located below H' since $h(a') < 0$.

EXAMPLE 2   Draw the graph of $y = \sin x + \sqrt{3} \cos x$ where $0 \leq x \leq 2\pi$.

SOLUTION   Let $g(x) = \sin x$ and $h(x) = \sqrt{3} \cos x$. Then $f(x) = g(x) + h(x)$. The result appears in Figure 4–40. The remarks in the solution of the previous example with respect to the points A, B, G, and H and $f(a) = g(a) + h(a)$ will apply here.

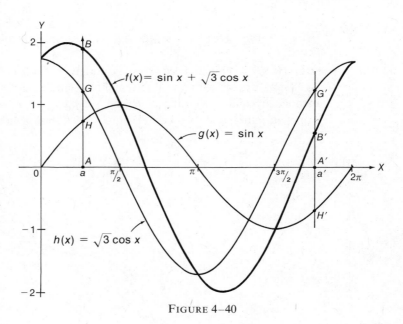

FIGURE 4–40

Note that the graph of $y = \sin x + \sqrt{3} \cos x$ looks very much like that of a sine curve which has been shifted to the left or a cosine curve shifted to the right. As a matter of fact it is, and another equation of the curve is $y = 2 \sin (x + \pi/3)$. If we had known this in the beginning, we could have sketched the graph comparatively rapidly. (You may already have done so. Refer to problem 5 in Section 4.2.) If $g(x)$ and $h(x)$ are sine and cosine curves with the same period, then $g(x) + h(x)$

4.6 ADDITION OF ORDINATES

is always a sine or cosine curve having that period. First observe that

$$2 \sin (x + \pi/3) = 2(\sin x \cos \pi/3 + \cos x \sin \pi/3)$$
$$= 2 \sin x \left(\frac{1}{2}\right) + 2 \cos x \left(\frac{1}{2}\sqrt{3}\right) = \sin x + \sqrt{3} \cos x$$

If we want to change $y = a \sin x + b \cos x$ to the form $y = A \sin (x + C)$, we can expand by means of the addition identity to obtain

$$A \sin (x + C) = A (\sin x \cos C + \cos x \sin C)$$
$$= A \sin x \cos C + A \cos x \sin C$$
$$= (A \cos C) \sin x + (A \sin C) \cos x$$

Now if $a \sin x + b \cos x = (A \cos C) \sin x + (A \sin C) \cos x$, it should be true that $A \cos C = a$ and $A \sin C = b$. If it *is* true, then we have $A^2 \cos^2 C = a^2$ and $A^2 \sin^2 C = b^2$, and if we add the left members and right members we get

$$A^2 \cos^2 C + A^2 \sin^2 C = a^2 + b^2$$
$$A^2(\cos^2 C + \sin^2 C) = a^2 + b^2$$
$$A^2 = a^2 + b^2$$
$$A = \sqrt{a^2 + b^2}$$

To find C, we use $A \cos C = a$, from which $\cos C = a/A$, and $A \sin C = b$, from which $\sin C = b/A$. We can then combine the results and write $\tan C = b/a$, or $C = \text{Arc tan} (b/a)$.

In the previous example we have $y = \sin x + \sqrt{3} \cos x$ so that $a = 1$ and $b = \sqrt{3}$. Thus $A = \sqrt{(1)^2 + (\sqrt{3})^2} = \sqrt{1 + 3} = \sqrt{4} = 2$, and $C = \text{Arc tan} (\sqrt{3}/1) = \pi/3$. Therefore $y = \sin x + \sqrt{3} \cos x$ is equivalent to $y = 2 \sin (x + \pi/3)$.

EXAMPLE 3   Change $y = 2 \sin x - 3 \cos x$ to the form $y = A \sin (x + C)$. Sketch the graph over the interval $0 \leq x \leq 2\pi$.

SOLUTION   Here $a = 2$ and $b = -3$, so $A = \sqrt{(2)^2 + (-3)^2} = \sqrt{4 + 9} = \sqrt{13}$, and $C = \text{Arc tan} (-3/2)$. Thus $y = \sqrt{13} \sin [x + \text{Arc tan} (-3/2)]$ is the required equation. To sketch the graph we use $\sqrt{13} \approx 3.61$, and $\text{Arc tan} (-3/2) \approx -0.98$. Again we revert to a decimal scale, and the result is shown in Figure 4–41. We can see that the maximum value occurs when $x + \text{Arc tan} (-3/2) = \pi/2$, or when $x = \pi/2 - \text{Arc tan} (-3/2) = \pi/2 + \text{Arc tan} (3/2) \approx 1.57 + 0.98 = 2.55$, the minimum occurs when $x + \text{Arc tan} (-3/2) = 3\pi/2$, or when $x \approx 4.71 + 0.98 = 4.69$. When $x = 0$, $y = 3.61 \sin [\text{Arc tan} (-3/2)] \approx 3.61 \sin (-0.98) \approx 3.61(-0.8305) \approx -3.00$.

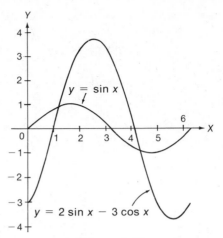

FIGURE 4–41

In the previous two examples we have assumed that the periods have all been the same. If the periods of the sine and cosine curves are different, then their sum is no longer a sine curve, but we can apply the method of addition of ordinates in order to draw the graph.

EXAMPLE 4 Sketch the graph of $f(x) = \sin x + \frac{1}{2} \cos 2x$ over the interval $0 \le x \le 2\pi$.

SOLUTION Draw $g(x) = \sin x$ and $h(x) = \frac{1}{2} \cos 2x$, then find points of $f(x) = g(x) + h(x)$ by addition of ordinates as before. The result will be something like that in Figure 4–42. By means of calculus it is possible to determine that the relative maximum or minimum values of this curve occur where $\cos x - \sin 2x = 0$, which motivates us to solve the equation over the interval $0 \le x \le 2\pi$. The solutions $x \in \{\pi/6, \pi/2, 5\pi/6, 3\pi/2\}$ seem to fit in nicely, since $f(\pi/6) = \sin \pi/6 + \frac{1}{2} \cos (2 \cdot \pi/6) = \frac{1}{2} + \frac{1}{2}(\frac{1}{2}) = 3/4$, $f(\pi/2) = \sin \pi/2 + \frac{1}{2} \cos (2 \cdot \pi/2) = 1 + \frac{1}{2}(-1) = \frac{1}{2}$, and so on.

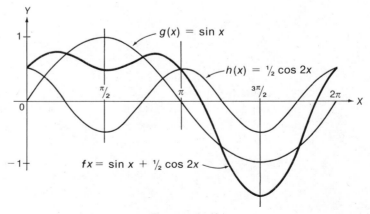

FIGURE 4–42

REVIEW EXERCISES

EXERCISES

Sketch the graph of each of the following over the interval $0 \leq x \leq 2\pi$, using the method of addition of ordinates.

1. $y = 1 + \cos x$
2. $y = 2 - \sin x$
3. $y = \sin x + \cos x$
4. $y = \cos x - \sqrt{3} \sin x$
5. $y = \sqrt{3} \sin x - \cos x$
6. $y = \sqrt{2} \sin x + \sqrt{2} \cos x$
7. $y = 3/2 \sin x + 2 \cos x$
8. $y = 3/2 \cos x + 2 \sin x$

Sketch the graph of each of the following over the interval indicated. Use the method of addition of ordinates.

9. $y = \cos x - x, \; -\pi \leq x \leq \pi$
10. $y = 2x - \cos x, \; -\pi \leq x \leq \pi$
11. $y = x^2 + \cos\left(\dfrac{\pi}{2} x\right), \; -2 \leq x \leq 2$
12. $y = x^2 - \sin \pi x, \; -2 \leq x \leq 2$
13. $y = \sin x + 1/2 \sin 2x, \; 0 \leq x \leq 2\pi$. *Hint: Maximum or minimum values may occur where $\cos 2x + \cos x = 0$.*
14. $y = \cos x - 1/2 \cos 2x, \; 0 \leq x \leq 2\pi$. *Hint: Maximum or minimum values may occur when $\sin 2x - \sin x = 0$.*
15. $y = \sqrt{3}/2 \sin 2x + 1/2 \cos 2x, 0 \leq x \leq 2\pi$
16. Show that the equation $y = m \cos x + n \sin x$ can be transformed into the form $y = A \cos (x - C)$ if $A = \sqrt{m^2 + n^2}$ and $C = \text{Arc tan}(n/m)$.
17. Transform the following equations into the form $y = A \sin B(x + C)$. Give the coordinates of the maximum and minimum points in the interval $0 \leq x \leq 2\pi$.
    a. $y = \sin x + \cos x$ *(refer to problem 3)*.
    b. $y = \sqrt{3} \sin x - \cos x$ *(refer to problem 5)*.
    c. $y = 3/2 \sin x + 2 \cos x$ *(refer to problem 7)*.
    d. $y = \sqrt{3}/2 \sin 2x + 1/2 \cos 2x$ *(refer to problem 15)*.
18. Transform the following equations into the form $y = A \cos B(x - C)$. Give the coordinates of the maximum and minimum points in the interval $0 \leq x \leq 2\pi$ (use the results of problem 16).
    a. $y = \cos x - \sqrt{3} \sin x$ *(refer to problem 4)*.
    b. $y = \sqrt{2} \sin x + \sqrt{2} \cos x$ *(refer to problem 6)*.
    c. $y = 3/2 \cos x + 2 \sin x$ *(refer to problem 8)*.
    d. $y = 1/2\sqrt{2} \cos 2x - 1/2\sqrt{2} \sin 2x$

REVIEW EXERCISES

Sketch the graph of each of the following over one complete cycle. Give the coordinates of zeros, relative maximum and minimum points, and equations of the asymptotes, if any.

1. $y = \sin 3x$
2. $y = \dfrac{1}{2} \tan 2x$
3. $y = \cos \dfrac{1}{2} x$
4. $y = \dfrac{4}{3} \cot \dfrac{3}{4} x$
5. $y = \dfrac{4}{5} \csc x$
6. $y = \dfrac{6}{5} \sec \dfrac{\pi}{2} x$
7. $y = \sin\left(x - \dfrac{\pi}{3}\right)$
8. $y = \dfrac{3}{5} \cos 2\left(x + \dfrac{\pi}{6}\right)$
9. $y = \tan \dfrac{\pi}{2}(x + 0.2)$

10. $y = \csc\left(x - \dfrac{\pi}{4}\right)$

Determine the exact value of each of the following:

11. Arc sin $(-1/2)$
12. Arc tan 1
13. Arc cos $(-\tfrac{1}{2}\sqrt{3})$
14. Arc cos (-1)
15. Arc tan (-1)
16. cos [Arc cos $(1/3)$]
17. csc [Arc sin $(5/9)$]
18. tan [Arc cos $(24/25)$]
19. sin [Arc cos $(-\sqrt{5}/3)$]
20. cos [Arc tan $(-\sqrt{15}/7)$]

Draw the graph of each of the following relations and their inverses, using a separate pair of axes for each relation-inverse pair. In each case tell whether the relation or its inverse is a function. Give the domain and range of each inverse.

21. $A = \{(x, y) \mid y = \dfrac{1}{4}x + 1 \text{ and } -4 \le x \le 4\}$
22. $B = \{(x, y) \mid y = \sqrt{x^2 - 1} \text{ and } 1 \le x \le 4\}$
23. $C = \{(x, y) \mid y = \dfrac{1}{2}\sqrt{16 - x^2}\}$
24. $D = \{(x, y) \mid y = \text{Cos } x\}$

Draw the graph of each of the following over the specified interval, using the method of addition of ordinates.

25. $y = \sqrt{3} \sin x - \cos x$ where $0 \le x \le 2\pi$
26. $y = \sqrt[3]{x} + \cos \dfrac{\pi}{2} x$ where $-8 \le x \le 8$

27. Transform the equation in problem 25 into the form $y = A \sin(x + C)$. Give the amplitude, period, and phase shift of the curve.

CHAPTER FIVE

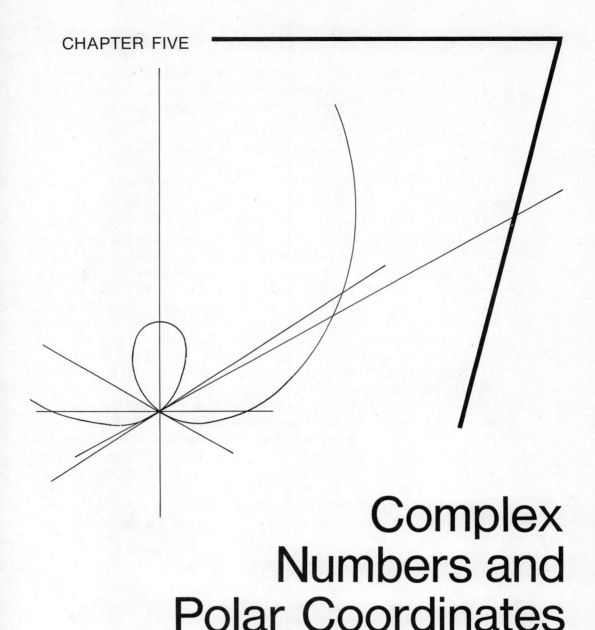

Complex Numbers and Polar Coordinates

5.1 COMPLEX NUMBERS IN ALGEBRAIC FORM

You have probably studied complex numbers in algebra. This section will serve as a review of some of the fundamental assumptions and operations with these numbers, which can be written in the form $a + bi$. In subsequent sections you will see how complex numbers can be written in a trigonometric form which facilitates the multiplication and division operations, and also makes finding powers and roots extremely simple.

In using the algebraic or rectangular form of a complex number, $a + bi$, we assume that a and b are real numbers and that i is defined as the number such that $i^2 = -1$, or $i = \sqrt{-1}$. We also assume the following definitions:

If a, b, c, and d are real numbers, then
EQUALITY $\qquad a + bi = c + di$ if and only if $a = c$ and $b = d$
ADDITION $\qquad (a + bi) + (c + di) = (a + c) + (b + d)i$
MULTIPLICATION $\quad (a + bi)(c + di) = (ac - bd) + (ad + bc)i$

With these definitions and using the corresponding axioms for real numbers, we can show that the set C of complex numbers obeys all the field axioms of the set of real numbers, that is, the closure, commutative, associative, identity, inverse, and distributive axioms. That the closure axiom is satisfied is evident from the definitions, since $a + c$ and $b + d$ are real numbers according to the closure axiom for addition of real numbers. Thus the sum of two complex numbers is a unique complex number. There is an identity element for addition, $0 + 0i$, since for any complex number $a + bi$, $(a + bi) + (0 + 0i) = (a + 0) + (b + 0)i = a + bi$ by the definition of addition for complex numbers and the identity axiom for addition of real numbers.

Although the definitions appear difficult to follow, in practice it is possible to treat complex numbers as algebraic expressions and to use the ordinary operations of the algebra of real numbers, replacing i^2 in the final result with -1, then simplifying. Conventionally we identify the real number a with the complex number $a + 0i$, and call the complex number $a + bi$, $b \neq 0$, an IMAGINARY NUMBER. If $a = 0$ and $b \neq 0$, then $0 + bi$ or bi is a PURE IMAGINARY NUMBER. It is sometimes convenient to refer to the numbers a and b of a complex number $a + bi$ as the *real part* and the *imaginary part*, respectively, even though both a and b are real numbers.

EXAMPLE 1 $\quad$ Perform the indicated operations and write the result in $a + bi$ form. $\quad$ a. $(5 + 3i) + (-2 + i)$ $\quad$ b. $(2 - 5i) - (3 - 4i)$
c. $(4 + 3i)^2$ $\quad$ d. $(3 + 5i)(2 - 3i)$

SOLUTIONS $\quad$ a. $(5 + 3i) + (-2 + i) = [5 + (-2)] + (3i + i) = 3 + 4i$
b. $(2 - 5i) - (3 - 4i) = 2 - 5i - 3 + 4i = (2 - 3) + (-5i + 4i) = -1 - i$
c. $(4 + 3i)^2 = 4^2 + 2(4)(3i) + (3i)^2 = 16 + 24i + 9i^2 = 16 + 24i + 9(-1) = 7 + 24i$
d. $(3 + 5i)(2 - 3i) = 6 - 9i + 10i - 15i^2 = 6 + i - 15(-1) = 21 + i$

The subtraction above was performed by the ordinary operations of algebra. We could recall the definition of subtraction for real numbers, $a - b = a + (-b)$,

5.1 COMPLEX NUMBERS IN ALGEBRAIC FORM

then show that $(a + bi) - (c + di) = (a - c) + (b - d)i$, since

$$(a + bi) - (c + di) = (a + bi) + [-(c + di)]$$
$$= (a + bi) + (-c - di)$$
$$= [a + (-c)] + [b + (-d)]i$$
$$= (a - c) + (b - d)i$$

The CONJUGATE of the complex number $a + bi$ is $a - bi$. The conjugate of $a - bi$ is $a + bi$. Since $(a + bi)(a - bi) = a^2 - b^2i^2 = a^2 - b^2(-1) = a^2 + b^2$, we can see that the product of a complex number and its conjugate is always a real number. This property is particularly useful in division of complex numbers. We could demonstrate that

$$\frac{a + bi}{c + di} = \frac{ac + bd}{c^2 + d^2} + \frac{bc - ad}{c^2 + d^2}i$$

but if we simply follow the practice of multiplying the numerator and denominator of the indicated quotient by the conjugate of the denominator, the results will be the same. For example,

$$\frac{3 + 2i}{5 - 3i} \cdot \frac{5 + 3i}{5 + 3i} = \frac{15 + 19i + 6i^2}{25 - 9i^2} = \frac{9 + 19i}{34} = \frac{9}{34} + \frac{19}{34}i$$

Since we plan to show in a later section how easy it is to find powers of complex numbers in trigonometric form, now might be a good time to show that it is difficult to find them in rectangular form. Finding the square of a complex number is not bad: $(a + bi)^2 = (a^2 - b^2) + 2abi$. The cube of $a + bi$ could then be obtained by multiplying $(a + bi)^2$ by $a + bi$, that is,

$$(a + bi)^2(a + bi) = [(a^2 - b^2) + 2abi] \cdot (a + bi)$$
$$= a^3 - ab^2 + 2a^2bi + a^2bi - b^3i + 2ab^2i^2$$
$$= (a^3 - 3ab^2) + (3a^2b - b^3)i$$

Just for the record $(a + bi)^5 = (a^5 - 10a^3b^2 + 5ab^4) + (5a^4b - 10a^2b^3 + b^5)i$, although of course with particular values of a and b the results will appear much simpler.

EXAMPLE 2 Find the indicated powers in $a + bi$ form: a. $(2 + 3i)^3$
b. $(1 - i)^4$ c. $\left(\frac{1}{2} - \frac{1}{2}\sqrt{3}i\right)^3$

SOLUTIONS   Using the binomial expansion,

$$(a+b)^n = a^n + n \cdot a^{n-1}b + \frac{n(n-1)}{1 \cdot 2} a^{n-2}b^2$$
$$+ \frac{n(n-1)(n-2)}{1 \cdot 2 \cdot 3} a^{n-3}b^3 + \cdots + b^n$$

we find

a. $(2 + 3i)^3 = 2^3 + 3(2)^2(3i) + 3(2)(3i)^2 + (3i)^3$
$= 8 + 36i + 54i^2 + 27i^3$
$= 8 + 36i + 54(-1) + 27(-1) \cdot i = -46 + 9i$

b. $(1 - i)^4 = 1^4 + 4(1)^3(-i) + 6(1)^2(-i)^2 + 4(1)(-i)^3 + (-i)^4$
$= 1 - 4i + 6i^2 - 4i^3 + i^4$
$= 1 - 4i + 6(-1) - 4(-1)i + (-1)(-1) = -4 + 0i$

c. $\left(\frac{1}{2} - \frac{1}{2}\sqrt{3}i\right)^3 = \left(\frac{1}{2}\right)^3 + 3\left(\frac{1}{2}\right)^2\left(-\frac{1}{2}\sqrt{3}i\right)$
$+ 3\left(\frac{1}{2}\right)\left(-\frac{1}{2}\sqrt{3}i\right)^2 + \left(-\frac{1}{2}\sqrt{3}i\right)^3$
$= \frac{1}{8} - \frac{3}{8}\sqrt{3}i + \frac{9}{8}i^2 - \frac{3}{8}\sqrt{3}i^3$
$= \frac{1}{8} - \frac{3}{8}\sqrt{3}i - \frac{9}{8} - \frac{3}{8}\sqrt{3}(-1)i = -1 + 0i$

We could first find

$$\left(\frac{1}{2} - \frac{1}{2}\sqrt{3}i\right)^2 = \frac{1}{4} - 2\left(\frac{1}{2}\right)\left(\frac{1}{2}\sqrt{3}i\right) + \left(-\frac{1}{2}\sqrt{3}i\right)^2$$
$$= \frac{1}{4} - \frac{1}{2}\sqrt{3}i + \frac{3}{4}i^2 = -\frac{1}{2} - \frac{1}{2}\sqrt{3}i$$

then

$$\left(\frac{1}{2} - \frac{1}{2}\sqrt{3}i\right)^3 = \left(\frac{1}{2} - \frac{1}{2}\sqrt{3}i\right)^2 \cdot \left(\frac{1}{2} - \frac{1}{2}\sqrt{3}i\right)$$
$$= \left(-\frac{1}{2} - \frac{1}{2}\sqrt{3}i\right)\left(\frac{1}{2} - \frac{1}{2}\sqrt{3}i\right)$$
$$= -\frac{1}{4} + \frac{3}{4}i^2 = -1 + 0i$$

5.1 COMPLEX NUMBERS IN ALGEBRAIC FORM

A cyclical property of i which is generally pointed out in algebra texts is useful in simplifying expressions such as those in the preceding examples. This involves powers of i, for which we observe that

$$i = i$$
$$i^2 = -1$$
$$i^3 = i^2 \cdot i = -i$$
$$i^4 = i^2 \cdot i^2 = (-1)(-1) = 1$$
$$i^5 = i^4 \cdot i = 1 \cdot i = i$$
$$i^6 = i^4 \cdot 1^2 = 1 \cdot i^2 = -1$$
$$i^7 = i^4 \cdot i^3 = 1 \cdot i^3 = -i$$
$$i^8 = i^4 \cdot i^4 = 1 \cdot 1 = 1$$
$$i^9 = i^8 \cdot i = 1 \cdot i = i$$

Any integral power of i can be changed to one of the numbers $-1, 1, -i$, or i.

EXAMPLE 3   Simplify: a. i^{15} b. i^{50} c. i^{88} d. i^{-3}

SOLUTIONS   a. $i^{15} = i^{12} \cdot i^3 = (i^4)^3 \cdot i^3 = 1^3 \cdot i^3 = 1 \cdot (-i) = -i$
b. $i^{50} = i^{48} \cdot i^2 = (i^4)^{12} \cdot i^2 = 1^{12} \cdot i^2 = 1 \cdot (-1) = -1$
c. $i^{88} = (i^4)^{22} = 1^{22} = 1$
d. $i^{-3} = 1 \cdot i^{-3} = i^4 \cdot i^{-3} = i^1 = i$ or $i^{-3} = \frac{1}{i^3} \cdot \frac{i}{i} = \frac{i}{i^4} = \frac{i}{1} = i$

We might have eliminated a step or two in each of the above had we first proved that $i^{4k} = (i^4)^k = 1^k = 1$, so that $i^{50} = i^{48} \cdot i^2 = i^{4(12)} \cdot i^2 = 1(-1) = -1$.

Another problem which is easier to solve when the number is in trigonometric form, but which is presented now for purposes of comparison, is that of finding the square roots of a complex number.

EXAMPLE 4   Find the square roots of $3 - 4i$.

SOLUTION   Let $x + yi$ represent the square root of $3 - 4i$. Then $(x + yi)^2 = 3 - 4i$, but $(x + yi)^2 = x^2 + 2xyi + y^2i^2 = (x^2 - y^2) + 2xyi$. If $(x^2 - y^2) + 2xyi = 3 - 4i$, then the *real parts*, $x^2 - y^2$ and 3, must be equal, and the *imaginary parts*, $2xy$ and -4, must be equal. Thus we have a system of equations

$$x^2 - y^2 = 3 \tag{1}$$

$$2xy = -4 \tag{2}$$

If we solve the second equation for y, $y = -2/x$. Substituting into the first equation:

$$x^2 - \left(-\frac{2}{x}\right)^2 = 3$$

$$x^2 - \frac{4}{x^2} = 3$$

$$x^4 - 4 = 3x^2$$

$$x^4 - 3x^2 - 4 = 0$$

Factoring, we obtain $(x^2 - 4)(x^2 + 1) = 0$, so that $x^2 - 4 = 0$ or $x^2 + 1 = 0$. The second equation has imaginary roots so we reject them, since x and y must be real. The first equation has 2 and -2 as roots, so if $x = 2$, $y = -1$ and if $x = -2$, $y = 1$ when we substitute into equation (2) above. Thus the two square roots of $3 - 4i$ are $2 - i$ and $-2 + i$, which can be verified by squaring each number.

EXERCISES

Perform each of the following operations and express the results in $a + bi$ form:

1. $(5 + 3i) + (-2 + 5i)$
2. $(3 - 7i) + (4 + 3i)$
3. $(6 - i) + (5 + 2i)$
4. $(-3 + 2i) + (1 - i)$
5. $(2 - 3i) - (5 - 3i)$
6. $(5 + 2i) - (5 - 2i)$
7. $(-3 + 4i) - (-2 + 3i)$
8. $(-5 - 3i) - (3 + 2i)$
9. $3(2 + 3i)$
10. $-2(3 - 4i)$
11. $-4(-1 + i)$
12. $5(-3 - 3i)$
13. $i(1 + i)$
14. $2i(3 - 2i)$
15. $-5i(2 - i)$
16. $-3i(-3 + 2i)$
17. $(5 + 2i)(1 - i)$
18. $(3 + 5i)(2 + 3i)$
19. $(4 - i)(-2 + 3i)$
20. $(-3 + i)(1 + 3i)$
21. $(3 + 4i)(3 - 4i)$
22. $(-5 + 3i)(-5 - 3i)$
23. $(2 - 5i)(2 + 5i)$
24. $(4 - i)(4 + i)$
25. $(a + bi)(a - bi)$
26. $(c - di)(c + di)$
27. $(2 + 5i)^2$
28. $(5 + 2i)^2$
29. $(3 - 4i)^2$
30. $(5 - 3i)^2$
31. $(x + yi)^2$
32. $(x - yi)^2$
33. $\dfrac{12 + 5i}{4}$
34. $\dfrac{-9 - 6i}{3}$
35. $\dfrac{-3 + 2i}{i}$
36. $\dfrac{5 - 5i}{i}$
37. $\dfrac{-3i}{2 + i}$
38. $\dfrac{4i}{1 - i}$
39. $\dfrac{-3 - 2i}{-3 + 2i}$
40. $\dfrac{-1 - 4i}{-1 + 4i}$
41. $\dfrac{3 - 2i}{4 + 3i}$
42. $\dfrac{1 + i}{3 - 3i}$
43. $\dfrac{4 + 2i}{2 + 3i}$
44. $\dfrac{3 + 5i}{5 - 3i}$

45. i^{13} 46. i^{27} 47. i^{30}
48. i^{53} 49. i^{-5} 50. i^{-6}
51. $(2+i)^3$ 52. $(1-i)^3$
53. $\left(\dfrac{1}{2}\sqrt{3}-\dfrac{1}{2}i\right)^3$ 54. $\left(\dfrac{1}{2}\sqrt{3}+\dfrac{1}{2}i\right)^3$
55. $(1+i)^6$
56. $(1-i)^6$

Find the square roots of each of the following in $a+bi$ form:

57. $3+4i$
58. $5-12i$
59. $-7+24i$
60. $-15-8i$

5.2 GRAPHICAL REPRESENTATION; TRIGONOMETRIC FORM OF COMPLEX NUMBERS

The set of real numbers, R, obeys certain properties of order, such as the TRICHOTOMY AXIOM: For any two real numbers a and b, exactly one of the following is true—$a<b$, $a=b$, or $a>b$. The complex numbers cannot be ordered, but they *can* be graphed if we define the point P with coordinates (a, b) to represent the complex number $a+bi$. We write $(a, b) = a+bi$, and the graph is simply that of Figure 5–1.

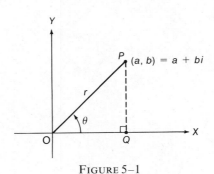

FIGURE 5–1

This is a somewhat familiar picture by now, and it is evident that $r = \sqrt{a^2+b^2}$, $\cos\theta = \dfrac{a}{r}$, and $\sin\theta = \dfrac{b}{r}$ so long as r is not zero. Since $a = r\cos\theta$ and $b = r\sin\theta$, we can write $a+bi = r\cos\theta + (r\sin\theta)i$, or in simpler form, we can factor to obtain

$$a+bi = r(\cos\theta + i\sin\theta)$$

The right member is the TRIGONOMETRIC FORM of the complex number. We can say that $r = \sqrt{a^2+b^2} = |a+bi|$ is the absolute value, or modulus, and that the angle θ in standard position such that $\tan\theta = b/a$ is the amplitude (or argument) of the number in its trigonometric form. It is easy to change from the rectangular $a+bi$ form to the trigonometric $r(\cos\theta + i\sin\theta)$ form or vice versa.

EXAMPLE 1   Change to trigonometric form
a. $3 + 3i$ b. $-\sqrt{3} + i$ c. $3 - 4i$

SOLUTIONS   It helps to draw the point on coordinate axes at first, as in Figure 5–2.   a. $r = \sqrt{(3)^2 + (3)^2} = \sqrt{9 + 9} = \sqrt{18} = 3\sqrt{2}$; $\tan \theta = 3/3 = 1$, so $\theta = 45°$. Thus $3 + 3i = 3\sqrt{2}(\cos 45° + i \sin 45°)$.   b. $r = \sqrt{(-\sqrt{3})^2 + (1)^2} = \sqrt{3 + 1} = \sqrt{4} = 2$; $\tan \theta = \dfrac{1}{-\sqrt{3}} = -\dfrac{1}{\sqrt{3}}$. $\theta = 30°$ and $\theta = 150°$ since $(-\sqrt{3}, 1)$ is in Quadrant II. Thus $-\sqrt{3} + i = 2(\cos 150° + i \sin 150°)$.   c. $r = \sqrt{(3)^2 + (-4)^2} = \sqrt{9 + 16} = \sqrt{25} = 5$; $\tan \theta = -4/3$, so $\theta =$ Arc $\tan (4/3) \approx 53°10'$, and $\theta \approx 360° - 53°10' = 306°50'$, so $3 - 4i = 5(\cos 306°50' + i \sin 306°50')$.

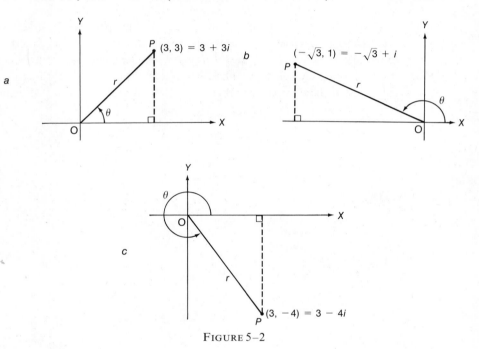

FIGURE 5–2

It may have occurred to you that the angle θ could really be any angle coterminal with those we have pictured. In the last example, for instance, we might have said $\theta = -53°10'$ and this would have been correct. It might have been a little awkward in this case to write $5[\cos(-53°10') + i \sin(-53°10')]$, but it would certainly locate the point for us.

EXAMPLE 2   Change to rectangular or algebraic form
a. $4(\cos 240° + i \sin 240°)$ b. $(\cos 90° + i \sin 90°)$
c. $2(\cos 335° + i \sin 335°)$

SOLUTIONS   It is still a good idea to draw the graph while becoming familiar with the relationships involved here. See Figure 5–3. First locate the angle, then use a convenient scale to determine the location of P, r

5.2 GRAPHICAL REPRESENTATION; TRIGONOMETRIC FORM OF COMPLEX NUMBERS

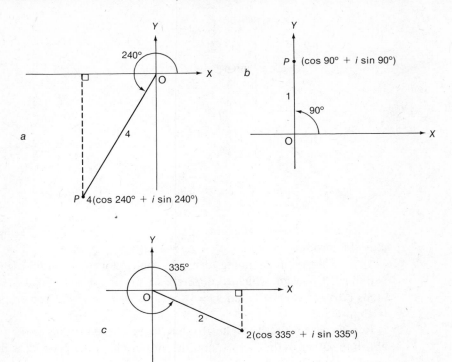

FIGURE 5-3

units from the origin. a. $\cos 240° = -\cos 60° = -1/2$, $\sin 240° = -\sin 60° = -\frac{1}{2}\sqrt{3}$, so $4(\cos 240° + i \sin 240°) = 4\cos 240° + (4\sin 240°)i = 4(-1/2) + 4(-\frac{1}{2}\sqrt{3})i = -2 - 2\sqrt{3}i$. b. $\cos 90° = 0$, $\sin 90° = 1$, $\cos 90° + i \sin 90° = 0 + i = i$. c. $\cos 335° = \cos 25° \approx 0.9063$, $\sin 335° = -\sin 25° \approx -0.4226$, so $2 \cos 335° + (2 \sin 335°)i \approx 2(0.9063) + 2(-0.4226)i = 1.8126 - 0.8452i$.

A common and convenient representation for a complex number is $z = a + bi$. We can write the conjugate of $a + bi$, or $a - bi$, as $\bar{z}$, and the reciprocal of $a + bi$ or $\frac{1}{(a + bi)}$, as z^{-1}. We have already demonstrated that the product of a complex number and its conjugate is always real, that is, $z \cdot \bar{z} = (a + bi)(a - bi) = a^2 - b^2i^2 = a^2 + b^2 = (a^2 + b^2) + 0i$. (What could you say about their sum?) We defined equality for two complex numbers in the previous section. In terms of z, if $z_1 = a_1 + b_1 i$ and $z_2 = a_2 + b_2 i$, then $z_1 = z_2$ if and only if $a_1 = a_2$ and $b_1 = b_2$. In trigonometric form, however, if $z_1 = r_1(\cos \theta_1 + i \sin \theta_1)$, and $z_2 = r_2(\cos \theta_2 + i \sin \theta_2)$, then $z_1 = z_2$ if and only if $r_1 = r_2$ and $\theta_2 = \theta_1 + k \cdot 360°$. Thus $3(\cos 20° + i \sin 20°) = 3(\cos 380° + i \sin 380°) = 3[\cos(-20°) + i \sin(-20°)]$.

To motivate our definition for multiplication of complex numbers in trigonometric form, let $z_1 = i$, $z_2 = -1 + i$, and then $z_3 = z_1 \cdot z_2 = i(-1 + i) = -i + i^2 = -1 - i$. If we graph these points, as in Figure 5-4, and change to trigonometric form, $z_1 = 1(\cos 90° + i \sin 90°)$, $z_2 = \sqrt{2}(\cos 135° + i \sin 135°)$ and $z_3 = \sqrt{2}(\cos 225° + i \sin 225°)$.

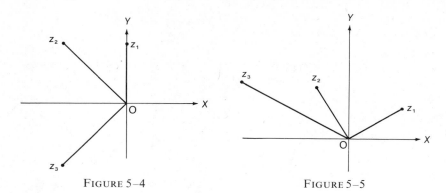

FIGURE 5-4   FIGURE 5-5

Selecting another two points, $z_1 = \sqrt{3} + i$, $z_2 = -1 + \sqrt{3}i$, we can calculate the product $z_3 = z_1 \cdot z_2 = (\sqrt{3} + i)(-1 + \sqrt{3}i) = -\sqrt{3} + 2i + \sqrt{3}i^2 = -2\sqrt{3} + 2i$. Again changing to trigonometric form and graphing (sometimes it is difficult to decide which to do first), we find $z_1 = 2(\cos 30° + i \sin 30°)$, $z_2 = 2(\cos 120° + i \sin 120°)$, and $z_3 = 4(\cos 150° + i \sin 150°)$. The graphs are shown in Figure 5-5.

In each case we see that the absolute value of the product is the product of the absolute values of the factors, and the amplitude of the product is the sum of the amplitudes of the factors. In symbolic form, if

$$z_1 = r_1(\cos \theta_1 + i \sin \theta_1)$$

and

$$z_2 = r_2(\cos \theta_2 + i \sin \theta_2)$$

then

$$z_3 = z_1 \cdot z_2 = r_1 r_2 [\cos (\theta_1 + \theta_2) + i \sin (\theta_1 + \theta_2)]$$

Thus if $z_1 = 3(\cos 50° + i \sin 50°)$ and $z_2 = 5(\cos 110° + i \sin 110°)$,

$$z_1 \cdot z_2 = 3 \cdot 5[\cos (50° + 110°) + i \sin (50° + 110°)]$$
$$= 15(\cos 160° + i \sin 160°)$$

It can also be shown that if $z_2 \neq 0$, then the quotient

$$\frac{z_1}{z_2} = \left(\frac{r_1}{r_2}\right) [\cos (\theta_1 - \theta_2) + i \sin (\theta_1 - \theta_2)]$$

or the absolute value of the quotient is the quotient of the absolute values and the amplitude of the quotient is the difference of the amplitudes of the dividend and divisor.

EXAMPLE 3   Given $z_1 = 6(\cos 130° + i \sin 130°)$ and $z_2 = 2(\cos 70° + i \sin 70°)$, find   a. $z_1 \cdot z_2$ b. $z_1 \div z_2$ c. z_1^2 d. $z_2 \div z_1$

5.2 GRAPHICAL REPRESENTATION; TRIGONOMETRIC FORM OF COMPLEX NUMBERS

SOLUTIONS

a. $z_1 \cdot z_2 = 6 \cdot 2[\cos(130° + 70°) + i\sin(130° + 70°)]$
$= 12(\cos 200° + i\sin 200°)$

b. $z_1 \div z_2 = \dfrac{6}{2}[\cos(130° - 70°) + i\sin(130° - 70°)]$
$= 3(\cos 60° + i\sin 60°)$

c. $z_1^2 = 6 \cdot 6[\cos(130° + 130°) + i\sin(130° + 130°)]$
$= 36(\cos 260° + i\sin 260°)$

d. $z_2 \div z_1 = \dfrac{2}{6}[\cos(70° - 130°) + i\sin(70° - 130°)]$
$= \dfrac{1}{3}[\cos(-60°) + i\sin(-60°)]$
$= \dfrac{1}{3}(\cos 300° + i\sin 300°)$

There is really nothing wrong with the negative angle in part d, but sometimes we may be asked to have the amplitude positive, or within a designated interval.

The method of multiplying and dividing using the trigonometric form is quite simple, but writing out $z = r(\cos \theta + i \sin \theta)$ is a bit cumbersome, so a common abbreviation for the trigonometric form is r cis θ. Then $3(\cos 50° + i \sin 50°) = 3$ cis $50°$, $10[\cos(-140°) + i \sin(-140°)] = 10$ cis $(-140°)$, and the definition of multiplication becomes

$$r_1 \text{ cis } \theta_1 \cdot r_2 \text{ cis } \theta_2 = r_1 r_2 \text{ cis } (\theta_1 + \theta_2)$$

EXERCISES

Locate each of the following on coordinate axes, then change to trigonometric form:

1. $\frac{1}{2} - \frac{1}{2}\sqrt{3}i$
2. $-1 - \sqrt{3}i$
3. $-2 + 2i$
4. $3\sqrt{2} - 3\sqrt{2}i$
5. $4\sqrt{3} - 4i$
6. $-4\sqrt{3} + 4i$
7. $-4 - 3i$
8. $5 - 12i$

Draw the graph of each of the following points then change it to rectangular form:

9. $5(\cos 180° + i \sin 180°)$
10. $4(\cos 270° + i \sin 270°)$
11. $6(\cos 120° + i \sin 120°)$
12. $5(\cos 330° + i \sin 330°)$
13. $4\sqrt{2}[\cos(-135°) + i \sin(-135°)]$
14. $4(\cos 945° + i \sin 945°)$
15. $5(\cos 220° + i \sin 220°)$
16. $6(\cos 100° + i \sin 100°)$

In problems 17 through 20 find a. $z_3 = z_1 + z_2$ b. $z_4 = z_1 - z_2$ c. $z_5 = z_1 \cdot z_2$ and draw the graph of each point.

17. $z_1 = i, z_2 = 1 + i$
18. $z_1 = i, z_2 = -2 + 3i$
19. $z_1 = 2 + i, z_2 = 3 - 2i$
20. $z_1 = -1 + 2i, z_2 = 2 - i$

In problems 21 through 24, find a. $z_1 \cdot z_2$ b. $z_1 \div z_2$ c. $z_2 \div z_1$ d. z_1^2 in trigonometric form.

21. $z_1 = 10(\cos 200° + i \sin 200°)$,
    $z_2 = 2(\cos 50° + i \sin 50°)$
22. $z_1 = 8(\cos 300° + i \sin 300°)$,
    $z_2 = 4(\cos 80° + i \sin 80°)$
23. $z_1 = 9(\cos 160° + i \sin 160°)$,
    $z_2 = 6[\cos(-40°) + i \sin(-40°)]$
24. $z_1 = 5[\cos(-20°) + i \sin(-20°)]$,
    $z_2 = 8[\cos(-120°) + i \sin(-120°)]$
25. Change to r cis θ form:
    a. $6(\cos 90° + i \sin 90°)$
    b. $10(\cos 170° + i \sin 170°)$
    c. $2[\cos(-15°) + i \sin(-15°)]$
    d. $3\sqrt{2} - 3\sqrt{2}i$
26. Write the definition for division of complex numbers which was given in trigonometric form in r cis θ form.

In problems 27 through 30, find a. $z_1 \cdot z_2$ b. $z_1 \div z_2$ c. $z_2 \div z_1$ d. z_1^2 in r cis θ form.

27. $z_1 = 8$ cis $60°$, $z_2 = 10$ cis $180°$
28. $z_1 = 4$ cis $310°$, $z_2 = 4$ cis $150°$
29. $z_1 = 6$ cis $(-30°)$, $z_2 = 4$ cis $(-90°)$
30. $z_1 = 10$ cis $(-50°)$, $z_2 = 5$ cis $390°$
31. Given $z = -3 - 4i$, find the square roots and draw the graphs of all three complex numbers.
32. Given $z = -5 + 12i$, find the square roots and draw the graphs of all three complex numbers.
33. Prove that the sum of any complex number and its conjugate is a real number.
34. Show that the reciprocal of $z = a + bi$ is $\dfrac{1}{z} = \dfrac{a}{a^2 + b^2} - \dfrac{b}{a^2 + b^2} i$.
35. Draw the graphs of $z = -3 + 4i$ and its reciprocal (refer to problem 34).
36. Draw the graphs of $z = 2(\cos 30° + i \sin 30°)$ and its reciprocal.
37. Show that $(\cos 0° + i \sin 0°)$ is the identity element for multiplication of complex numbers in trigonometric form. *Hint:* In the algebra of real numbers, 1 is the identity element for multiplication since $a \cdot 1 = a$ for every real number a.
38. Show that $\dfrac{1}{r}[\cos(-\theta) + i \sin(-\theta)]$ is the inverse element for multiplication of complex numbers in trigonometric form. *Hint:* In the algebra of real numbers, $1/a$ is the inverse element for multiplication since $a \cdot (1/a) = 1$ for all real numbers a. (The product of any number and its multiplicative inverse is the multiplicative identity. Refer to problem 37.)
39. Prove that if $z_1 = r_1(\cos \theta_1 + i \sin \theta_1)$ and $z_2 = r_2(\cos \theta_2 + i \sin \theta_2)$, then $z_3 = z_1 \cdot z_2 = r_1 r_2 [\cos(\theta_1 + \theta_2) + i \sin(\theta_1 + \theta_2)]$. *Hint: Multiply and see what happens.*

5.3 POWERS AND ROOTS OF COMPLEX NUMBERS

Our experience with multiplication of complex numbers in trigonometric form should encourage us with respect to raising a given number to natural number

5.3 POWERS AND ROOTS OF COMPLEX NUMBERS

powers. We can see that if $z = r \operatorname{cis} \theta$, then

$$z^2 = r \operatorname{cis} \theta \cdot r \operatorname{cis} \theta = r \cdot r \operatorname{cis} (\theta + \theta) = r^2 \operatorname{cis} 2\theta$$

$$z^3 = z \cdot z^2 = r \operatorname{cis} \theta \cdot r^2 \operatorname{cis} 2\theta = r \cdot r^2 \operatorname{cis} (\theta + 2\theta) = r^3 \operatorname{cis} 3\theta$$

$$z^4 = z \cdot z^3 = r \operatorname{cis} \theta \cdot r^3 \operatorname{cis} 3\theta = r \cdot r^3 \operatorname{cis} (\theta + 3\theta) = r^4 \operatorname{cis} 4\theta$$

$$\cdots$$

$$z^n = z \cdot z^{n-1} = r \operatorname{cis} \theta \cdot r^{n-1} \operatorname{cis} [(n-1)\theta]$$

$$= r \cdot r^{n-1} \operatorname{cis} [\theta + (n-1)\theta] = r^n \operatorname{cis} n\theta \qquad (1)$$

The result

$$z^n = (r \operatorname{cis} \theta)^n = r^n \operatorname{cis} n\theta$$

for n a natural number is known as DE MOIVRE'S THEOREM. If the number is already in trigonometric form, the simplification is easy, that is, $(2 \operatorname{cis} 100°)^5 = 2^5 \operatorname{cis} (5 \cdot 100°) = 32 \operatorname{cis} 500° = 32 \operatorname{cis} 140°$.

EXAMPLE 1   Find the value of each of the following in rectangular form:

a. $(1 - i)^4$ b. $\left(\dfrac{1}{2} - \dfrac{1}{2}\sqrt{3}i\right)^3$ c. $\dfrac{(1 + i)^6}{(\frac{1}{2}\sqrt{3} - \frac{1}{2}i)^3}$ d. $(2 + 3i)^3$

SOLUTIONS

a. $(1 - i)^4 = (\sqrt{2} \operatorname{cis} 315°)^4 = (2^{1/2})^4 \operatorname{cis} (4 \cdot 315°)$

$\qquad = 2^2 \operatorname{cis} 1260° = 4 \operatorname{cis} 180° = 4(\cos 180° + i \sin 180°)$

$\qquad = 4(-1 + 0i) = -4 + 0i$

b. $(\frac{1}{2} - \frac{1}{2}\sqrt{3}i)^3 = (\operatorname{cis} 300°)^3 = 1^3 \operatorname{cis} (3 \cdot 300°)$

$\qquad = \operatorname{cis} 900° = \operatorname{cis} 180°$

$\qquad = \cos 180° + i \sin 180° = -1 + 0i$

c. $\dfrac{(1 + i)^6}{(\frac{1}{2}\sqrt{3} - \frac{1}{2}i)^3} = \dfrac{(\sqrt{2} \operatorname{cis} 45°)^6}{(\operatorname{cis} 330°)^3} = \dfrac{(2^{1/2})^6 \operatorname{cis} (6 \cdot 45°)}{1^3 \operatorname{cis} (3 \cdot 330°)}$

$\qquad = \dfrac{2^3 \operatorname{cis} 270°}{\operatorname{cis} 990°} = \dfrac{8 \operatorname{cis} 270°}{\operatorname{cis} (-90°)}$

$\qquad = 8 \operatorname{cis} [270° - (-90°)] = 8 \operatorname{cis} 360°$

$\qquad = 8(\cos 360° + i \sin 360°) = 8(1 + 0i)$

$\qquad = 8 + 0i$

d. $(2 + 3i)^3 \approx (\sqrt{13} \operatorname{cis} 56°20')^3 = (13^{1/2})^3 \operatorname{cis} (3 \cdot 56°20')$

$\qquad = 13^{3/2} \operatorname{cis} 169° = 13\sqrt{13}(\cos 169° + i \sin 169°)$

$\qquad \approx 13\sqrt{13}(-0.9816 + 0.1908i) \approx -46 + 8.94i$

To be consistent with the algebra of real numbers it is necessary to define $z^0 = 1 = 1 + 0i = 1 \text{ cis } 0°$, and $z^{-n} = \dfrac{1}{z^n}$. In each case z must not be the zero element, which would be $z = 0 = 0 + 0i = 0 \text{ cis } \theta$. (In defining the trigonometric form we assumed $r \neq 0$, but if $r = 0$, then θ can be any angle.) Thus if $z = 3 \text{ cis } 50°$, we would find that

$$z^2 \cdot z^0 = (3 \text{ cis } 50°)^2 \cdot 1 \text{ cis } 0° = 9 \text{ cis } 100° \cdot 1 \text{ cis } 0°$$
$$= 9 \cdot 1 \text{ cis }(100° + 0°) = 9 \text{ cis } 100° = (3 \text{ cis } 50°)^2 = z^2$$

Also

$$z^3 \cdot z^{-2} = (3 \text{ cis } 50°)^3 \cdot \dfrac{1}{(3 \text{ cis } 50°)^2} = 3^3 \text{ cis }(3 \cdot 50°) \cdot \dfrac{\text{cis } 0°}{3^2 \text{ cis }(2 \cdot 50°)}$$
$$= 27 \text{ cis } 150° \cdot \dfrac{1 \text{ cis } 0°}{9 \text{ cis } 100°} = 27 \text{ cis } 150° \cdot \dfrac{1}{9} \text{ cis }(-100°)$$
$$= 27 \cdot \dfrac{1}{9} \text{ cis }[150° + (-100°)] = 3 \text{ cis } 50° = z$$

Note that we didn't assume $z^3 \cdot z^{-2} = z^{3+(-2)} = z^1 = z$, which would be the case if z were a real number, but rather worked through with the definitions of multiplication and division which we have at our disposal. The results are consistent.

One of the interesting applications of the trigonometric form of complex numbers and De Moivre's Theorem involves finding roots of complex numbers. If $(z_k)^n = z$, we shall call z_k an nth root of z. Thus $2 \text{ cis } 30°$ is a fourth root of $16 \text{ cis } 120°$, since $(2 \text{ cis } 30°)^4 = 2^4 \text{ cis }(4 \cdot 30°) = 16 \text{ cis } 120°$. *Finding* the fourth roots of $16 \text{ cis } 120°$ is another matter. Let us examine the procedure first in this special case. Suppose that $z_k = t \text{ cis } \phi$ is a fourth root of $16 \text{ cis } 120°$. Then

$$(t \text{ cis } \phi)^4 = t^4 \text{ cis } 4\phi = 16 \text{ cis } 120°$$

But if $t^4 \text{ cis } 4\phi = 16 \text{ cis } 120°$, then

$$t^4 = 16 \text{ and } 4\phi = 120° + k \cdot 360°$$

according to the definition of equality for complex numbers in trigonometric form.

Now t is a positive real number, so if $t^4 = 16$ then t must equal 2, not -2, nor $\pm 2i$, even though all three of the latter are fourth roots of 16. If $4\phi = 120° + k \cdot 360°$, then $\phi = 30° + k \cdot 90°$. Thus $\phi \in \{30°, 120°, 210°, 300°, 390°, 480°, \ldots\}$ when $k = 0$, $k = 1$, $k = 2$, $k = 3$, $k = 4$, $k = 5$, and so on. We have already verified that $z_0{}^4 = (2 \text{ cis } 30°)^4 = 16 \text{ cis } 120°$. (We use the symbol z_0 to mean the root obtained

5.3 POWERS AND ROOTS OF COMPLEX NUMBERS

when $k = 0$.) Next we check

$$z_1^4 = (2 \text{ cis } 120°)^4 = 2^4 \text{ cis } (4 \cdot 120°) = 16 \text{ cis } 480° = 16 \text{ cis } 120°$$
$$z_2^4 = (2 \text{ cis } 210°)^4 = 2^4 \text{ cis } (4 \cdot 210°) = 16 \text{ cis } 840° = 16 \text{ cis } 120°$$
$$z_3^4 = (2 \text{ cis } 300°)^4 = 2^4 \text{ cis } (4 \cdot 300°) = 16 \text{ cis } 1200° = 16 \text{ cis } 120°$$
$$z_4^4 = (2 \text{ cis } 390°)^4 = 2^4 \text{ cis } (4 \cdot 390°) = 16 \text{ cis } 1560° = 16 \text{ cis } 120°$$

After z_3, however, we are merely repeating ourselves since $z_4 = z_0$, $z_5 = z_1$, and so on. The graphs of z and its 4 fourth roots are shown in Figure 5-6.

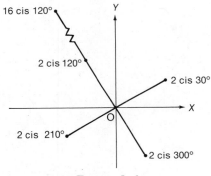

FIGURE 5-6

If we generalize in order to find the nth root of any complex number $r \text{ cis } \theta$, then if

$$(z_k)^n = (t \text{ cis } \phi)^n = r \text{ cis } \theta = z$$
$$(t \text{ cis } \phi)^n = t^n \text{ cis } n\phi = r \text{ cis } \theta,$$

so $\qquad t^n = r \qquad$ and $\qquad n\phi = \theta + k \cdot 360°$

Solving the first equation for the positive root of r and the second equation for ϕ, we obtain

$$t = r^{1/n} = \sqrt[n]{r} \qquad \text{and} \qquad \phi = \left(\frac{\theta}{n}\right) + k \cdot \left(\frac{360°}{n}\right)$$

Thus

$$z_k = r^{1/n} \text{ cis}\left(\frac{\theta}{n} + k \cdot \frac{360°}{n}\right), \quad k \in \{0, 1, 2, 3, \ldots, n - 1\}. \tag{2}$$

Note that we stop at $k = n - 1$, because $k = n$ would repeat a result already found, and there are n roots since $k = 0$ is counted.

EXAMPLE 2   Find the cube roots of 1 and plot the results.

SOLUTION   The real number $1 = 1 + 0i = \text{cis } 0°$, so substituting $n = 3$, $r = 1$, $\theta = 0°$ in equation (2):

$$z_k = 1^{1/3} \text{ cis} \left(\frac{0°}{3} + k \cdot \frac{360°}{3} \right) = \text{cis } (k \cdot 120°), \quad k \in \{0, 1, 2\}$$

Then $z_0 = \text{cis } 0°$, $z_1 = \text{cis } 120°$, $z_2 = \text{cis } 240°$. Unless stated otherwise, we would generally change the results back to rectangular form again, so $z_0 = 1$, $z_1 = -\frac{1}{2} + \frac{1}{2}\sqrt{3}i$, and $z_2 = -\frac{1}{2} - \frac{1}{2}\sqrt{3}i$. The graph is shown in Figure 5–7.

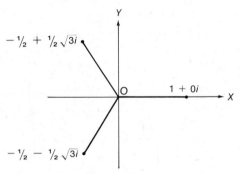

FIGURE 5–7

EXAMPLE 3   Find the fifth roots of $32 \text{ cis } 300°$ and plot the results.

SOLUTION   Substituting $n = 5$, $r = 32$, $\theta = 300°$ in equation (2)

$$z_k = 32^{1/5} \text{ cis} \left(\frac{300°}{5} + k \cdot \frac{360°}{5} \right) = 2 \text{ cis } (60° + k \cdot 72°), \quad k \in \{0, 1, 2, 3, 4\},$$

so $z_0 = 2 \text{ cis } 60°$, $z_1 = 2 \text{ cis } 132°$, $z_2 = 2 \text{ cis } 204°$, $z_3 = 2 \text{ cis } 276°$, $z_4 = 2 \text{ cis } 348°$. The graph is shown in Figure 5–8.

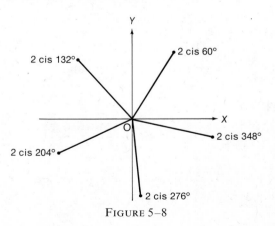

FIGURE 5–8

5.3 POWERS AND ROOTS OF COMPLEX NUMBERS

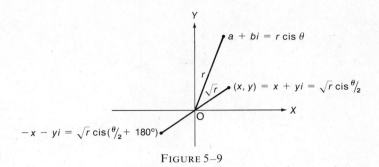

FIGURE 5-9

In Section 5.1 we found the square roots of $3 - 4i$ by letting $(x + yi)^2 = 3 - 4i$ and equating the real and imaginary parts, which resulted in a system of two equations in two variables. Using the results of De Moivre's Theorem can make this operation simpler. Referring to Figure 5-9, we see that $x^2 + y^2 = r = \sqrt{a^2 + b^2}$. Since $(x + yi)^2 = (x^2 - y^2) + 2xyi = a + bi$, we also have $x^2 - y^2 = a$ and $2xy = b$. Now we have three equations in two variables. Applying this to finding the square roots of $3 - 4i$, $r = \sqrt{3^2 + (-4)^2} = \sqrt{9 + 16} = \sqrt{25} = 5$, so

$$x^2 + y^2 = 5 \quad (1) \qquad 2xy = -4 \text{ or } y = -\frac{2}{x} \quad (3)$$

$$\underline{x^2 - y^2 = 3 \quad (2)}$$

Solving the first two equations for x by addition we obtain $x = \pm 2$. Substituting $x = 2$ into equation (3) gives $y = -1$, and substitution of $x = -2$ gives $y = 1$ so that the two square roots are $2 - i$ and $-2 + i$.

EXAMPLE 4 Find the square roots of $-15 + 8i$ in rectangular form and graph the results.

SOLUTION $a = -15, b = 8$, and $r = \sqrt{(-15)^2 + (8)^2} = \sqrt{225 + 64} = \sqrt{289} = 17$, so

$$x^2 + y^2 = 17 \quad (1) \qquad 2xy = 8 \text{ or } y = \frac{4}{x} \quad (3)$$

$$\underline{x^2 - y^2 = -15 \quad (2)}$$
$$2x^2 = 2$$

Adding (1) and (2), we get $2x^2 = 2$, $x^2 = 1$, so $x = \pm 1$. From equation (3) we obtain $y = 4$ upon substituting $x = 1$, and if $x = -1$, $y = -4$. The graph is shown in Figure 5-10.

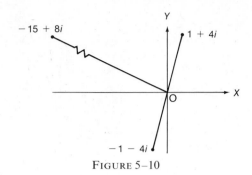

FIGURE 5-10

EXERCISES

Find the value of each of the following in trigonometric form:

1. $(3 \text{ cis } 50°)^4$
2. $(5 \text{ cis } 100°)^3$
3. $(\sqrt{2} \text{ cis } 110°)^6$
4. $(\sqrt{3} \text{ cis } 210°)^4$
5. $(-1 + i)^4$
6. $(\sqrt{2} - \sqrt{2}i)^4$
7. $(\sqrt{3} + i)^6$
8. $(-1 - \sqrt{3}i)^5$
9. $\dfrac{(-\sqrt{3} + i)^4}{(1 - i)^8}$
10. $\dfrac{(2 - 2\sqrt{3}i)^3(-2\sqrt{2} + 2\sqrt{2}i)^5}{(4\sqrt{3} - 4i)^2}$

Find the indicated roots in trigonometric form and plot the results.

11. The cube roots of 8 cis 60°
12. The fourth roots of cis 100°
13. The sixth roots of 8 cis 0°
14. The eighth roots of 16 cis 0°
15. The square roots of i
16. The cube roots of -1

Find the indicated roots in rectangular form and plot the results.

17. The fourth roots of 1
18. The fourth roots of -1
19. The cube roots of -1
20. The cube roots of i
21. The cube roots of $4\sqrt{2} + 4\sqrt{2}i$
22. The fourth roots of $-8 + 8\sqrt{3}i$

Find the square roots of each of the following in rectangular form.

23. $-5 - 12i$
24. $7 - 24i$

25. Find the sum of the roots in each of the problems 17, 19, and 21.

26. If $n > 2$ the consecutive nth roots of a number, when joined by line segments, form what kind of figure?

5.4 POLAR COORDINATES

The trigonometric form of a complex number is closely related to the concept of polar coordinates of a point. If we have a point P, not at the origin, with rectangular coordinates (x, y) as shown in Figure 5-11, then if θ is the angle formed by OP and the x axis, and if ρ (Greek letter rho) is the distance of P from the origin,

5.4 POLAR COORDINATES

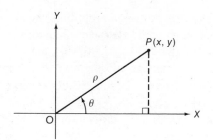

FIGURE 5–11

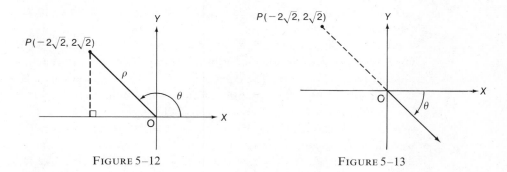

FIGURE 5–12 FIGURE 5–13

$\cos \theta = x/\rho$ and $\sin \theta = y/\rho$, so that $x = \rho \cos \theta$ and $y = \rho \sin \theta$. We also observe that $\rho^2 = x^2 + y^2$ and $\tan \theta = y/x$.

The notation (ρ, θ) with ρ and θ related as above will be called the POLAR COORDINATES of the point P. One reason for using ρ rather than r is that we have previously insisted that r be positive. If $\rho^2 = x^2 + y^2$, then $\rho = \pm\sqrt{x^2 + y^2}$, so we are permitting ρ to be either positive or negative, although we shall see that (ρ, θ) and $(-\rho, \theta)$ are two different points in the plane. We define the polar coordinates of the origin to be $(0, \theta)$, with θ any angle.

We can change from rectangular to polar coordinate form and also from polar to rectangular form. For example, if we have the point P with rectangular coordinates $(-2\sqrt{2}, 2\sqrt{2})$, as shown in Figure 5–12, $\rho^2 = (-2\sqrt{2})^2 + (2\sqrt{2})^2 = 8 + 8 = 16$, so $\rho = \pm 4$, and $\tan \theta = 2\sqrt{2}/(-2\sqrt{2}) = -1$, so $\theta = 45°$ and θ as indicated in the figure would be 135°. In order to locate a point whose polar coordinates are given, we may think in terms of standing at the origin, facing in the direction 0°, then rotating through the angle θ. If ρ is positive we move out ρ units in the direction we are facing, if ρ is negative we move *back* $|\rho|$ units in the opposite direction. So (4, 135°) would be one set of polar coordinates for P, and (4, 135° + $k \cdot$ 360°) would represent an infinite set of possible polar coordinates for P. We could use (4, −225°) or (4, 495°) or any one of the many multiples of 360° combined with 135° to represent θ, although if we used $k \geq 6$ the idea of our doing the rotating through the angle θ might not work out too well.

For the above example we could also use $\theta = -45°$, since $\tan(-45°) = -1$, but then we would have to use $\rho = -4$ so that $(-4, -45°)$ or $(-4, -45° + k \cdot 360°)$ would be another infinite set of polar coordinates for P. Figure 5–13 shows $\theta = -45°$, and the dashed line OP is our method of indicating that ρ is negative.

We can see from this example that each point can have an infinite number of polar coordinates. However, each pair of polar coordinates represents a unique point. The points (4, 30°), (2, −240°), and (−3, 100°) are shown in Figure 5–14.

In order to illustrate the method of changing from polar to rectangular coordinate form we can use the points in Figure 5–14. For (4, 30°) we have $x = 4 \cos 30° = 4(\frac{1}{2}\sqrt{3}) = 2\sqrt{3}$ and $y = 4 \sin 30° = 4(\frac{1}{2}) = 2$, so $(2\sqrt{3}, 2)$ are the rectangular coordinates. To change (2, −240°) we use $x = 2 \cos(−240°) = 2(−\cos 60°) = 2(−1/2) = −1$ and $y = 2 \sin(−240°) = 2 \sin 60° = 2(\frac{1}{2}\sqrt{3}) = \sqrt{3}$, so $(−1, \sqrt{3})$ are the rectangular coordinates. For the third point, (−3, 100°), $x = −3 \cos 100° \approx −3(−\cos 80°) \approx −3(−0.1736) \approx .52$, and $y = −3 \sin 100° = −3 \sin 80° \approx −3(0.9848) \approx −2.95$. All of these values seem to be confirmed by the figure.

One of the advantages of the polar coordinate system is that some equations which are fairly complicated in terms of the variables x and y are relatively simple in polar coordinate form. The equation $x^2 + y^2 = 16$ becomes $\rho^2 = 16$ or $\rho = \pm 4$ in polar form. The latter is the set of all points 4 units from the origin (θ can be any angle), which would of course be a circle, and that is exactly what the graph of $x^2 + y^2 = 16$ would be. The equation $y = x$ could be changed to $\rho \sin \theta = \rho \cos \theta$, or $\sin \theta / \cos \theta = \tan \theta = 1$, so $\theta = 45°$ would be a polar equation of that line. Note that ρ could be any number, positive or negative. We have to agree that the origin, (0, 45°), would also be a point on the line.

The transition from one type of equation to the other is not always simple, however, and we shall not spend too much time on that type of exercise. The situation is somewhat analogous to that of changing from the English to the metric system of measurement and vice versa. The changes can be made, but we usually operate within one system or the other.

Graphing polar equations can be a tedious process, but the results are sometimes quite interesting. To draw the graph of $\rho = \sin 3\theta$, for example, we can make a table of values for θ and ρ. The graph and the table of values are shown in Figure 5–15. Note that after we reach 180° the points begin to duplicate each other. Plotting more points in between, however, would fill in the gaps and we can sketch a smooth curve as shown in the figure. It will turn out that the graph

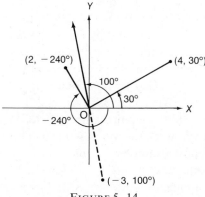

FIGURE 5–14

5.4 POLAR COORDINATES

θ	0°	15°	30°	45°	60°	75°	90°	105°
3θ	0°	45°	90°	135°	180°	225°	270°	315°
$\rho = \sin 3\theta$	0	0.71	1.00	0.71	0	−0.71	−1.00	−0.71

θ	120°	135°	150°	165°	180°	195°	210°	225°
3θ	360°	405°	450°	495°	540°	585°	630°	675°
$\rho = \sin 3\theta$	0	0.71	1.00	0.71	0	−0.71	−1.00	−0.71

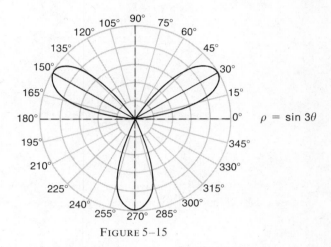

FIGURE 5–15

of $\rho = \sin n\theta$ (n a natural number) will be a petal curve, in which there are n petals if n is odd and $2n$ petals if n is even.

EXAMPLE 1   Draw the graph of $\rho = 2 + 3 \cos \theta$.

SOLUTION   It is best to start at $\theta = 0°$ and find values at regular intervals. A 15° interval is usually adequate until you become more familiar with the appearance of polar curves. This interval has the advantage of involving the special angles, although tables are usually necessary for 15°, 75°, and so on. If $f(\theta) = 2 + 3 \cos \theta$, then

$$f(0°) = 2 + 3 \cos 0° = 2 + 3(1) = 2 + 3 = 5$$

$$f(15°) = 2 + 3 \cos 15° \approx 2 + 3(0.9659) \approx 2 + 2.9 = 4.9$$

$$f(30°) = 2 + 3 \cos 30° = 2 + 3 \cdot \frac{1}{2}\sqrt{3} = 2 + 1.5\sqrt{3} \approx 4.6$$

or $f(30°) \approx 2 + 3(0.8660) \approx 2 + 2.6 = 4.6$

Continuing in the same way we obtain the polar coordinates (4.1, 45°), (3.5, 60°), (2.8, 75°), (2, 90°), (1.2, 105°), (.5, 120°), (−.1, 135°), (−.6, 150°),

$(-.9, 165°)$, $(-1, 180°)$, $(-.9, 195°)$, $(-.6, 210°)$, $(-.1, 225°)$, $(.5, 240°)$, $(1.2, 255°)$, $(2, 270°)$, $(2.8, 285°)$, $(3.5, 300°)$, $(4.1, 315°)$, $(4.6, 330°)$, $(4.9, 345°)$, and $(5, 360°)$. Plotting these points and connecting them with a smooth curve results in the graph of Figure 5–16.

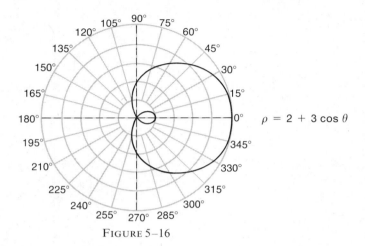

FIGURE 5–16

It is evident from the above that after 180° the points are related to the ones previously calculated by symmetry with respect to the x axis. It can be proved that

1. if replacement of (ρ, θ) by $(\rho, -\theta)$ results in the same equation, the graph is symmetric with respect to the x axis;

2. if replacement of (ρ, θ) by $(\rho, \pi - \theta)$ results in the same equation, the graph is symmetric with respect to the y axis;

3. if replacement of (ρ, θ) by either $(-\rho, \theta)$ or $(\rho, \pi + \theta)$ results in the same equation, the graph is symmetric with respect to the origin.

In the above example, $\rho = 2 + 3 \cos(-\theta) = 2 + 3 \cos \theta$, so condition 1 holds as was demonstrated.

Finding the points of intersection of two polar curves presents an interesting example of the need for solving trigonometric equations.

EXAMPLE 2   Draw the graphs of $\rho = 2\sqrt{2}$ and $\rho = 4 \cos 2\theta$. Give the coordinates of their points of intersection.

SOLUTION   The graphs are shown in Figure 5–17. Since $4 \cos[2(-\theta)] = 4 \cos(-2\theta) = 4 \cos 2\theta$, $4 \cos[2(\pi - \theta)] = 4 \cos(2\pi - 2\theta) = 4 \cos(-2\theta) = 4 \cos 2\theta$, and $4 \cos[2(\pi + \theta)] = 4 \cos(2\pi + 2\theta) = 4 \cos 2\theta$, all three of the conditions mentioned above hold and we see that each point of the graph is symmetric to another with respect to the x axis, the y axis, and also the origin. Replacing ρ with $4 \cos 2\theta$ in the first equation, we obtain $4 \cos 2\theta =$

5.4 POLAR COORDINATES

$2\sqrt{2}$, so $\cos 2\theta = \frac{1}{2}\sqrt{2}$. Thus $2\theta = 45°$ or $2\theta = 405°$ so that $\theta = 22\frac{1}{2}°$ or $\theta = 202\frac{1}{2}°$, and $2\theta = 315°$ or $2\theta = 675°$ so that $\theta = 157\frac{1}{2}°$ or $\theta = 337\frac{1}{2}°$. But there are eight points of intersection, so how can we find them? If we let $\rho = -2\sqrt{2}$ we also get the circle of Figure 5–17, therefore $4\cos 2\theta = -2\sqrt{2}$ means $\cos 2\theta = -\frac{1}{2}\sqrt{2}$. Then $2\theta = 135°$ or $2\theta = 395°$ so that $\theta = 67\frac{1}{2}°$ or $\theta = 197\frac{1}{2}°$, and $2\theta = 225°$ or $2\theta = 585°$ so that $\theta = 112\frac{1}{2}°$ or $\theta = 292\frac{1}{2}°$. This example is not intended to be discouraging, but rather to show the advantages of both a graphical and an analytical solution of the system.

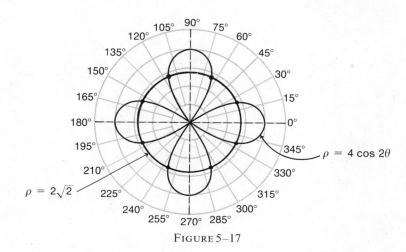

FIGURE 5–17

EXERCISES

Find polar coordinates for each of the following points if their rectangular coordinates are given.

1. $(0, 3)$
2. $(-4, 0)$
3. $(2, -2)$
4. $(-2\sqrt{3}, -2)$
5. $(-3, 4)$
6. $(24, -7)$

Find rectangular coordinates for each of the following points if their polar coordinates are given.

7. $(4, 270°)$
8. $(-2, 90°)$
9. $(6, 120°)$
10. $(4\sqrt{2}, 135°)$
11. $(10, 200°)$
12. $(5, 340°)$
13. $(4, 11\pi/6)$
14. $(6, 3)$

Change each equation to an equivalent equation in polar form. In each case solve for ρ.

15. $x^2 + y^2 - y = 0$
16. $x^2 + y^2 - x = 0$
17. $x = 4$
18. $y = -2$
19. $x^2 = 4 + 4y$
20. $3x^2 + 4y^2 - 6x = 9$

Change each equation to an equivalent equation in x and y. Eliminate all radical expressions. Assume $\sqrt{x^2 + y^2} \neq 0$.

21. $\rho = 3$
22. $\rho = 2\sin\theta$
23. $\rho = 2\cos\theta$
24. $\text{Arc} \tan \frac{1}{3} = \theta$
25. $\rho = \dfrac{2}{1 - \cos\theta}$
26. $\rho(1 - 2\cos\theta) = 4$

Graph each of the following for $0° \leq \theta \leq 360°$.

27. $\rho = 4 \cos \theta$
28. $\rho = 4 \sin (\theta + 45°)$
29. $\rho = 4 \sin 2\theta$
30. $\rho = 4 \cos 3\theta$
31. $\rho = 2 + 2 \sin \theta$
32. $\rho = 3 + \sin \theta$
33. $\rho = 3 - 2 \cos \theta$
34. $\rho = 2 - 3 \cos \theta$
35. $\theta = \text{Arc tan } \dfrac{1}{2}$
36. $\theta = \text{Arc tan } 3$
37. $\rho = 4\sqrt{\sin \theta}$
38. $\rho^2 = 16 \cos \theta$

Draw the graphs of each of the following pairs of equations on the same polar axes, then find their points of intersection analytically.

39. $\rho = 2$ and $\rho = 4 \cos \theta$
40. $\rho = 4 \sin \theta$ and $\rho = 4 \cos \theta$
41. $\rho = 3 \sin \theta$ and $\rho = 2 - \sin \theta$
42. $\rho = \sqrt{2 \cos \theta}$ and $\rho = 2(1 - \cos \theta)$

REVIEW EXERCISES

Perform each of the following operations and express the results in $a + bi$ form:

1. $(7 - 3i) + (-5 + 4i)$
2. $(-3 + 5i) + (4 - 3i)$
3. $(2 + i) - (2 - 4i)$
4. $(-3 + 3i) - (-2 + i)$
5. $(4 + 3i)(2 - 3i)$
6. $(-1 - 4i)(5 - 2i)$
7. $(-2 + i)^2$
8. $(4 - 3i)^2$
9. $\dfrac{2 - 3i}{3 - i}$
10. $\dfrac{-4 + 3i}{2 + 3i}$
11. i^{39}
12. i^{53}
13. $(3 + 2i)^3$
14. $(1 - 2i)^3$
15. $(1 - i)^4$
16. $(-1 + i)^5$
17. $\dfrac{(-1 + i)^4 (2 + 2i)^3}{(2\sqrt{3} - 2i)^3}$
18. $\dfrac{(\frac{1}{2} - \frac{1}{2}\sqrt{3}i)^5 (\sqrt{3} + i)^4}{(4 - 4i)^3}$

19. Find the square roots of $15 - 8i$ in $a + bi$ form.
20. Find the square roots of $-7 - 24i$ in $a + bi$ form.

Change to $a + bi$ form:

21. $6(\cos 150° + i \sin 150°)$
22. $3\sqrt{2}(\cos 315° + i \sin 315°)$
23. $4 \text{ cis } 225°$
24. $2 \text{ cis } 240°$
25. If $z_1 = 12(\cos 180° + i \sin 180°)$ and $z_2 = 4(\cos 300° + i \sin 300°)$, find
   a. $z_1 \cdot z_2$ b. $z_1 \div z_2$
   c. z_1^2 d. z_2^2
26. If $z_1 = 6(\cos 110° + i \sin 110°)$ and $z_2 = 3(\cos 230° + i \sin 230°)$, find
   a. $z_1 \cdot z_2$ b. $z_1 \div z_2$
   c. z_1^2 d. z_2^2
27. Find the fifth roots of $32 \text{ cis } 300°$.
28. Find the cube roots of $8i$.

29. Change $y^2 = 9 - 6x$ to an equivalent equation in polar form and solve for ρ.

30. Change $x^2 + y^2 + 4x = 0$ to an equivalent equation in polar form and solve for ρ.

31. Change $\rho = \dfrac{4}{2 + \sin \theta}$ to an equivalent equation in x and y. Eliminate any radical expressions.

32. Change $\rho = \dfrac{1}{1 + \cos \theta}$ to an equivalent equation in x and y. Eliminate any radical expressions.

Graph each of the following for $0° \leq \theta \leq 360°$:

33. $\rho = 5 \sin 3\theta$

34. $\rho = 5 \cos 2\theta$

35. $\rho = 2 \sin \theta - 3$

36. $\rho = 2(\cos \theta + 1)$

37. $\rho = \dfrac{2}{1 - \cos \theta}$

38. $\rho = \dfrac{4}{2 + \sin \theta}$

CHAPTER SIX

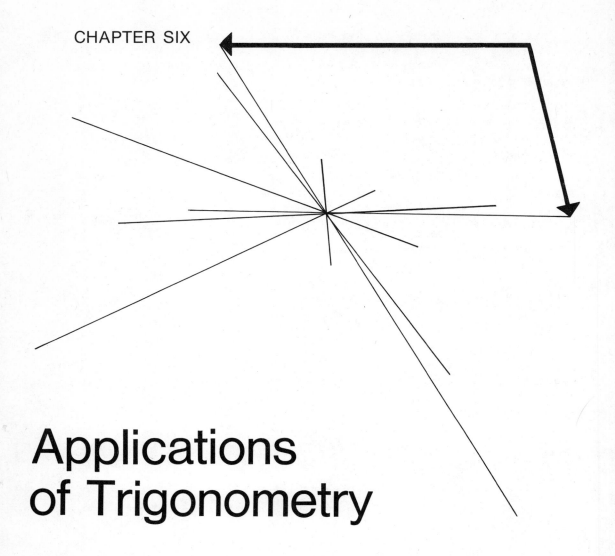

Applications of Trigonometry

6.1 THE LAW OF SINES

In Section 1.2 you were introduced to methods of solutions for right triangles and in Section 1.3 you were promised a wide range of practical applications. Although right triangle techniques will produce solutions for almost any triangle problem, there are more direct methods which are often simpler and more efficient. As a review of right triangle methods which will lead us into the Law of Sines, let us consider a problem of finding the area of a triangle.

Suppose we know that two sides of a triangle are 10 inches and 12 inches long, and that the angle included between them measures 40°. Can we find the area of the triangle? It usually helps to draw a figure such as that in Figure 6–1 in order to decide how to proceed. Labeling the figure is also helpful because it permits us to refer to significant parts of the figure as we work through the solution. Most of you will recall that in order to find the area of a triangle we can use the formula $\mathscr{A} = \frac{1}{2}bh$, in which b is the length of the base and h is the

6.1 THE LAW OF SINES

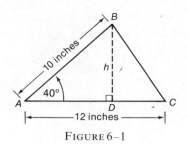

FIGURE 6-1

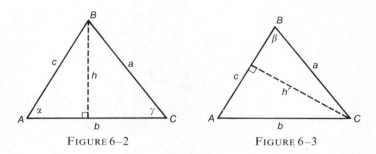

FIGURE 6-2   FIGURE 6-3

length of the altitude. If $b = AC$ then we have the base, but we don't know h, which is indicated by the dashed line segment BD in the figure. We can *find* h, however, since triangle ABD is a right triangle and $\sin 40° = h/10$, or $b = 10 \sin 40° = 10(.6428) = 6.428$. Thus

$$\mathscr{A} = \frac{1}{2}bh = \frac{1}{2}(12)(6.428) = 38.568 \approx 38.6 \text{ square inches}$$

The same procedure could be used to find the area of any triangle when the lengths of two sides and the measures of their included angle are known, but it is useful to derive a formula for solving the problem. Suppose we have a triangle ABC as shown in Figure 6-2. Its area can be found by means of the formula $\mathscr{A} = \frac{1}{2}bh$, and if we know the values of α and c, $\sin \alpha = \dfrac{h}{c}$ so that $h = c \sin \alpha$. If we know the values of γ and a, then $\sin \gamma = \dfrac{h}{a}$ and $h = a \sin \gamma$. Thus $\mathscr{A} = \frac{1}{2}bh = \frac{1}{2}bc \sin \alpha$ or $\frac{1}{2}ba \sin \gamma$. In either case, the area is one half the product of the lengths of two sides times the sine of their included angle. This should hold for a, c, and β, so we verify this by drawing an altitude from C to AB, as in Figure 6-3, which will be denoted by h'. The area of the triangle by the basic formula is one half the product of a side times the altitude to that side, so $\mathscr{A} = \frac{1}{2}ch'$, and since $\sin \beta = \dfrac{h'}{a}$ or $h' = a \sin \beta$, $\mathscr{A} = \frac{1}{2}ca \sin \beta$ as expected. In each case shown, the angle has been acute. We shall leave the proof for the case in which the angle is obtuse to the reader in the exercises.

EXAMPLE 1   Find the area of a triangle if $a = 20$, $c = 15$, and $\beta = 55°$.

SOLUTION We are assuming the standard labeling of the triangle, in which a is the length of the side opposite the angle with vertex at A and having measure α, and so on. Thus we have two sides and an included angle. $\mathscr{A} = \frac{1}{2}ac \sin \beta = \frac{1}{2}(20)(15) \sin 55° = 150 \sin 55°$, or $\mathscr{A} \approx 150(0.8192) = 122.88 \approx 123$. If there is any possibility of confusion it is certainly worthwhile to draw a sketch of the triangle and to label the given parts, as in Figure 6–4.

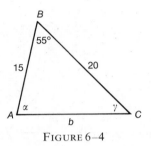

FIGURE 6–4

Now it is time for a further word concerning accuracy of calculations. The reader will note that the answer in the above example was stated in two ways— $150 \sin 55°$ and 123. Assuming the given measurements were exact, the first answer, $150 \sin 55°$, was also exact. The answer 123 was an approximation to three significant digits. As soon as we use the tables and start rounding off results, we cease to be exact. We would more than likely confound a clerk in a store if we went in to order enough paint to cover $150 \sin 55°$ square feet of signboard, however. The whole subject of precision, accuracy, rounding off numbers and significant digits, together with a discussion of the relative merits of using or not using calculators, slide rules, or logarithms in class, in working homework assignments, and in taking tests, is a matter to be discussed among students and teachers. We shall simply state our results in exact form whenever possible, followed by an approximation using the tables in this text and a calculator with at least an 8-digit capacity, as stated in Chapter One.

In Chapter One we solved right triangles such as that in Figure 6–5. That is, given $\gamma = 90°$, $\alpha = 20°$, and $c = 25$, we found β, a and b by means of geometry and trigonometric ratios. The measure of β is simply $90° - 20° = 70°$, $\sin 20° = a/25$ so that $a = 25 \sin 20° \approx 25(0.3420) \approx 8.55$, and $\cos 20° = b/25$ so that $b = 25 \cos 20° \approx 25(0.9397) \approx 23.5$.

Now suppose that we have a triangle which is not a right triangle, with $\gamma = 60°$, $\alpha = 20°$, and $b = 40$ as shown in Figure 6–6. The measure of $\beta =$

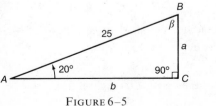

FIGURE 6–5

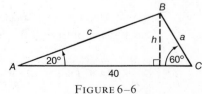

FIGURE 6–6

6.1 THE LAW OF SINES

$180° - (20° + 60°) = 100°$. Using the formula developed in Section 1.3, we could find

$$h = \frac{b}{\cot \alpha + \cot \gamma} = \frac{40}{\cot 20° + \cot 60°}$$

Then $\csc 20° = c/h$, so

$$c = h \csc 20° = \frac{40 \csc 20°}{\cot 20° + \cot 60°} \approx \frac{40(2.924)}{2.747 + 0.5774} = \frac{116.96}{3.3244} \approx 35.2$$

and $\csc 60° = \frac{a}{h}$, so

$$a = h \csc 60° = \frac{40 \csc 60°}{\cot 20° + \cot 60°} \approx \frac{40(1.155)}{2.747 + 0.5774} = \frac{46.2}{3.3244} \approx 13.9$$

This solution is certainly possible, but most would consider it complicated. At this point humans generally look for a better, simpler way to handle a problem of this type, which is where the Law of Sines can help. There are numerous ways to develop the relationship which is called the Law of Sines, but we can use the area formula developed earlier, that is,

$$\mathscr{A} = \frac{1}{2} bc \sin \alpha = \frac{1}{2} ac \sin \beta = \frac{1}{2} ab \sin \gamma$$

(The area remains constant, no matter how we calculate it.) Since no side of a triangle would have length equal to zero, we can divide each of the three right hand members by $\frac{1}{2}abc$, so that

$$\frac{\frac{1}{2}bc \sin \alpha}{\frac{1}{2}abc} = \frac{\frac{1}{2}ac \sin \beta}{\frac{1}{2}abc} = \frac{\frac{1}{2}ab \sin \gamma}{\frac{1}{2}abc}$$

and if we simplify each fraction we obtain

$$\frac{\sin \alpha}{a} = \frac{\sin \beta}{b} = \frac{\sin \gamma}{c}$$

This rather nice result is called the LAW OF SINES. In words, *the sines of the angles of any triangle are proportional to their opposite sides.*

If we apply this to the triangle of Figure 6–6, we find

$$\frac{\sin 20°}{a} = \frac{\sin 100°}{40} = \frac{\sin 60°}{c}$$

so that

$$a = \frac{40 \sin 20°}{\sin 100°} \approx \frac{40(0.3420)}{0.9848} \approx 13.9$$

150   SIX | APPLICATIONS OF TRIGONOMETRY

and
$$c = \frac{40 \sin 60°}{\sin 100°} \approx \frac{40(0.8660)}{0.9848} \approx 35.2$$

Did you notice that the results of the calculations correspond precisely with those found previously? It would be an interesting exercise to show that each of the equations

$$\frac{40 \sin 20°}{\sin 100°} = \frac{40 \csc 60°}{\cot 20° + \cot 60°}$$

and
$$\frac{40 \sin 60°}{\sin 100°} = \frac{40 \csc 20°}{\cot 20° + \cot 60°}$$

are indeed exactly equal without referring to tables.

EXAMPLE 2   Solve the triangle which has $a = 55$, $b = 60$ and $\beta = 47°20'$.

SOLUTION   Sketch a figure and label the parts, as in Figure 6–7. Then by the Law of Sines we have

$$\frac{\sin \alpha}{55} = \frac{\sin 47°20'}{60} = \frac{\sin \gamma}{c} \quad \text{and} \quad \sin \alpha = \frac{55 \sin 47°20'}{60}$$

so that

$$\alpha = \text{Arc sin} \left(\frac{11 \sin 47°20'}{12} \right)$$

and since $\alpha + \beta + \gamma = 180°$, $\gamma = 180° - (\alpha + \beta)$, while $c = \dfrac{60 \sin \gamma}{\sin 47°20'}$. The approximate values, then, are

$$\alpha \approx \text{Arc sin} \left[\frac{11(0.7353)}{12} \right] \approx \text{Arc sin } 0.6740 \approx 42°20'$$

$$\gamma \approx 180° - (47°20' + 42°20') = 90°20'$$

$$c = \frac{60 \sin 90°20'}{\sin 47°20'} \approx \frac{60(1.0000)}{0.7353} \approx 81.6$$

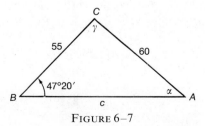

FIGURE 6–7

6.1 THE LAW OF SINES

EXAMPLE 3 An observation post at point A is 2000 yards west of an observation post at point B. Lookouts at each post simultaneously sight an object at point C which is 32° to the north of A and 41° north of B as indicated in Figure 6–8. What is the distance from A to C?

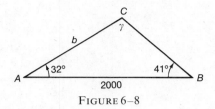

FIGURE 6–8

SOLUTION $\gamma = 180° - (\alpha + \beta) = 180° - (32° + 41°) = 180° - 73° = 107°$, and $\dfrac{\sin 41°}{b} = \dfrac{\sin \gamma}{2000}$ so

$$b = \frac{2000 \sin 41°}{\sin \gamma} = \frac{2000 \sin 41°}{\sin 107°} \approx \frac{2000(0.6561)}{0.9563} \approx 1370 \text{ yards}$$

EXERCISES

Solve each of the following triangles and also determine its area.

1. $c = 200, \alpha = 35°, \beta = 72°$
2. $a = 110, \beta = 25°, \gamma = 83°$
3. $b = 270, \alpha = 33°10', \gamma = 104°50'$
4. $c = 3.45, \alpha = 123°20', \gamma = 10°10'$
5. $a = 30, b = 40, \beta = 40°$
6. $b = 15.0, c = 25.0, \gamma = 100°$
7. $a = 5.23, c = 7.07, \gamma = 90°$
8. $a = 22.7, c = 50.0, \alpha = 27°00'$

9. Two ships are sailing due north when lookouts observe an object in the water between them but bearing 155° from the leading ship and 035° from the second ship. If the ships are 5000 yards apart, how far is the object from the second ship? (Refer to Figure 6–9.)

10. A surveyor determines the distance from A to B across a lake by measuring a line from A to C, 3000 feet long, and the angles BAC and ACB which measure 40° and 115°, respectively. What is the distance from A to B? (Refer to Figure 6–10.)

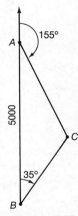

FIGURE 6–9

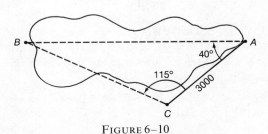

FIGURE 6–10

11. A helicopter flies 50 miles from A to B, then turns and flies 10 miles from B to C. If $\alpha = 8°20'$ and ABC is an obtuse angle, find the distance from C back to A. (*Refer to Figure 6–11.*)

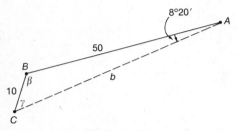

FIGURE 6–11

12. An airplane flies from an airport at point A to a city at point B, 270 miles away, then turns right $50°$ and heads for another city at point C, which is located 300 miles from A. What is the distance from B to C? (*Refer to Figure 6–12.*)

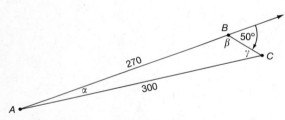

FIGURE 6–12

13. Show that $\dfrac{40 \csc 60°}{\cot 20° + \cot 60°}$ is exactly equal to $\dfrac{40 \sin 20°}{\sin 100°}$. (*Do not use tables.*)

14. Prove that in any triangle ABC, $\dfrac{b \csc \gamma}{\cot \alpha + \cot \gamma} = \dfrac{b \sin \alpha}{\sin \beta}$.

15. a. Solve the triangle ABC with $\alpha = 35°$, $a = 10$, and $c = 12$, calculating the angles to the nearest ten minutes and the sides to three significant digits.

b. The solution for part a should be $\gamma = 43°30'$, $\beta = 101°30'$, and $b = 17.1$. Prepare an accurate scale drawing of the triangle, using 1/4 inch = 1 unit.

c. Could angle ACB have been obtuse? In other words, could $\gamma' = 136°30'$? If so, find β' and b' and label the triangle which was drawn in part b appropriately.

16. In problem 11 it was assumed that the helicopter changed course on its flight from B to C so that angle ABC was an obtuse angle. Could B have been acute if it had not been stated otherwise? If this is so, find the distance from C back to A.

17. Can you solve the triangle with $\alpha = 40°$, $b = 20$ and $c = 15$ by a direct application of the Law of Sines? Demonstrate the reason for your answer.

18. Solve the triangle in problem 17 by means of right triangles. *Hint: Draw the triangle with its altitude from B to AC.*

19. To make an accurate drawing of a triangle with a given angle of measure α, the side opposite α, of length a, and the side with length c, do the following:

1. Use a protractor to draw the angle with measure α at point A

2. measure the length c from A to point B

3. using drawing compasses with points set a units apart, place one point at B and draw an arc which intersects the other side of angle A at point C.

Refer to Figure 6–13. Use $\alpha = 30°$, $c = 4''$, and answer the following questions:

a. If $a = 1$ inch, will the arc drawn from B intersect AX?

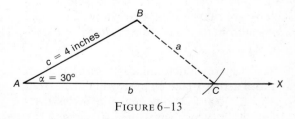

FIGURE 6–13

b. What happens when $a = BC = 2$ inches? (Perhaps the question is what *should* happen?)

c. Draw the triangle when $a = BC = 3$ inches. Could there be *two* such triangles?

d. Draw the triangle when $a = BC = 5$ inches.

20. Show that $\mathcal{A} = \frac{1}{2}bc \sin \alpha$ if $90° < \alpha < 180°$. (*Draw the triangle.*)

6.2 THE LAW OF SINES: AMBIGUOUS CASE

The exercises in the previous section illustrated some limitations of the Law of Sines—there may be no direct application that leads to a solution, there may be no solution at all, or there may be two distinct solutions. The cases in which the Law of Sines is directly applicable include those which are abbreviated SAA, ASA, and SSA, referring to *a side and two angles*, *two angles and their included side*, and *two sides and the angle opposite one of them*, respectively.

EXAMPLE 1 Solve the triangle for which $c = 4$, $a = 3$, and $\alpha = 30°$.

SOLUTION A sketch of the triangle is shown in Figure 6–14. Since $\frac{\sin 30°}{3} = \frac{\sin \gamma}{4}$,

$$\sin \gamma = \frac{4 \sin 30°}{3} = \frac{4(.5)}{3} = \frac{2}{3} \approx 0.6667$$

and
$$\gamma \approx \text{Arc sin } (0.6667) \approx 41°50'$$

Thus $\beta = 180° - (30° + \gamma) = 150° - \gamma \approx 150° - 41°50' = 108°10'$, and since $\frac{\sin 30°}{3} = \frac{\sin \beta}{b}$, $b \approx \frac{3(0.9502)}{0.5000} \approx 5.70$. These values seem reasonable when compared to the sketch of Figure 6–14.

There is another possibility, however, in that if $\gamma \approx 41°50'$, we can have $\gamma' = 180° - \gamma \approx 138°10'$, and then $\beta' = 180° - (30° + \gamma') = 150° - \gamma' \approx$

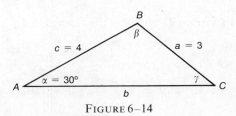

FIGURE 6–14

$150° - 138°10' = 11°50'$, while

$$b = \frac{3 \sin 11°50'}{\sin 30°} \approx \frac{3(0.2051)}{0.5000} \approx 1.23$$

These values are indicated in Figure 6–15.

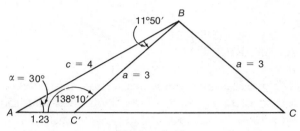

FIGURE 6–15

The situation in which we know two sides of a triangle and the angle opposite one of them (SSA) is called the AMBIGUOUS CASE, since there may be more than one solution. As a matter of fact there may be two, one, or no solutions depending upon the relative measures of sides and angles.

EXAMPLE 2   Solve the triangle for which $c = 4$, $a = 1$, and $\alpha = 30°$.

SOLUTION   Since $\dfrac{\sin 30°}{1} = \dfrac{\sin \gamma}{4}$, $\sin \gamma = \dfrac{4 \sin 30°}{1} = 4(.5) = 2$, but Arc sin 2 does not exist, so there is no solution for this problem. Such a triangle is not possible. Refer to Figure 6–16. The circle with center at B and radius 1 unit will not intersect the other side of angle A.

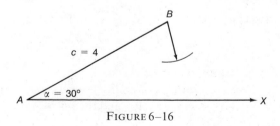

FIGURE 6–16

EXAMPLE 3   Solve the triangle for which $c = 4$, $a = 2$, and $\alpha = 30°$.

SOLUTION   Since $\dfrac{\sin 30°}{2} = \dfrac{\sin \gamma}{4}$, $\sin \gamma = \dfrac{4 \sin 30°}{2} = \dfrac{4(.5)}{2} = \dfrac{2}{2} = 1$, and $\gamma = $ Arc sin $1 = 90°$. Then $\beta = 180° - (30° + 90°) = 180° - 120° = 60°$,

6.2 THE LAW OF SINES: AMBIGUOUS CASE   155

and since $\dfrac{\sin 30°}{2} = \dfrac{\sin \beta}{b}$, $b = \dfrac{2 \sin 60°}{\sin 30°} = 4 \sin 60° \approx 4(0.8660) \approx 3.46$.

Refer to Figure 6–17. The circle with center at B and radius 2 units is tangent to the other side of angle A, so angle ACB is a right triangle.

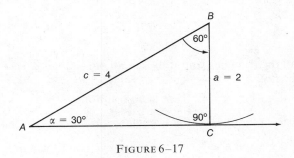

FIGURE 6–17

EXAMPLE 4   Solve the triangle for which $c = 4$, $a = 5$, and $\alpha = 30°$.

SOLUTION   Since $\dfrac{\sin 30°}{5} = \dfrac{\sin \gamma}{4}$, $\sin \gamma = \dfrac{4 \sin 30°}{5} = \dfrac{4(.5)}{5} = \dfrac{2}{5} = 0.4000$, and $\gamma = \text{Arc} \sin 0.4 \approx 23°30'$. Then $\beta = 180° - (30° + \gamma) = 150° - \gamma \approx 150° - 23°30' = 126°30'$, and since $\dfrac{\sin 30°}{5} = \dfrac{\sin \beta}{b}$, $b \approx \dfrac{5 \sin 126°30'}{\sin 30°} = 10 \sin 53°30' \approx 10(0.8039) \approx 8.04$.

Could there be another solution for this triangle, as was the case in the first example? Suppose $\gamma' \approx 23°30'$, so that $\gamma' \approx 180° - 23°30' = 156°30'$. Then $\beta' = 180° - (30° + \gamma') = 150° - \gamma' \approx 150° - 156°30' = -6°30'$, which is not possible. Referring to Figure 6–18 it can be seen that the circle with center at B and radius 5 intersects the other side of angle A at point C, but that C' is beyond the endpoint of ray AC, so will not intersect it.

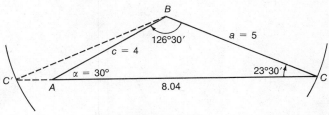

FIGURE 6–18

For a given value of c and α, then, we have observed the possibilities of two solutions, no solution, and one solution, depending upon the relative lengths of c and a. All are summarized in Figure 6–19. Here we see that BC_3 is the altitude

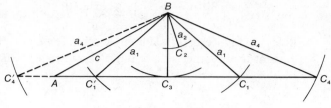

FIGURE 6–19

of all the triangles, and if the length of BC_3 is $h = c \sin \alpha$ units, then

1 when $a < h$, there is no solution;
2 when $a = h$, the triangle is a right triangle and there is one solution;
3 when $h < a < c$ there are two solutions; and
4 when $h < a$ and $a > c$ there is one solution.

We are also assuming that $\alpha < 90°$. If $\alpha \geq 90°$, then there is no triangle if $a \leq c$ and only one if $a > c$.

The labeling here is arbitrary, and you need to remember the relationship between the side opposite the given angle and the altitude of the triangle, and whether or not the side opposite the given angle is less than or greater than the other given side.

EXAMPLE 5   How many solutions are possible if $b = 10$, $c = 12$, and $\gamma = 65°$?

SOLUTION   Sketch the possible triangle as shown in Figure 6–20. Although not required it might be a good idea always to locate the given angle and the given side not opposite the given angle in the same relative positions. Then it is easy to see that since $c > b$, there is but one solution.

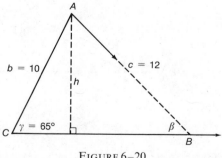

FIGURE 6–20

EXAMPLE 6   How many solutions are possible if $a = 10$, $b = 5$, and $\beta = 72°$?

SOLUTION   Sketch the possible triangle as shown in Figure 6–21. Since $\sin 72° = \frac{h}{10}$ and $h = 10 \sin 72° \approx 10(.9511) \approx 9.51$, $b < h$, and there is no solution.

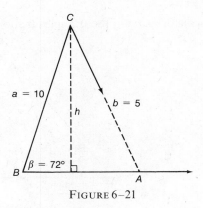

FIGURE 6–21

EXAMPLE 7   Solve the triangle for which $c = 12$, $b = 10$, and $\beta = 50°$.

SOLUTION   Sketch the possible triangle as shown in Figure 6–22. Since $\sin 50° = \frac{h}{12}$ and $h = 12 \sin 50° \approx 12(0.7660) \approx 9.19$, we see that $h < b < c$ and there must be two solutions. To find the first, use $\frac{\sin 50°}{10} = \frac{\sin \gamma}{12}$, so that $\sin \gamma = \frac{12 \sin 50°}{10} = 1.2 \sin 50° \approx 1.2(0.7660) \approx 0.9190$. Then $\gamma =$ Arc sin $(1.2 \sin 50°) \approx 66°50'$, so $\alpha = 180° - (50° + \gamma) = 130° - \gamma \approx 130° - 66°50' = 63°10'$, and $\frac{\sin \alpha}{a} = \frac{\sin 50°}{10}$ so that

$$a = \frac{10 \sin \alpha}{\sin 50°} \approx \frac{10(0.8923)}{0.7660} \approx 11.6$$

To find the second solution let $\gamma =$ Arc sin $(1.2 \sin 50°)$, so that $\gamma' =$

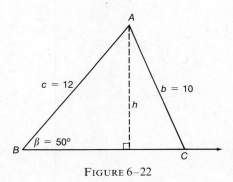

FIGURE 6–22

$180° - \gamma \approx 180° - 66°50' = 113°10'$. Then $\alpha' = 180° - (50° + \gamma') = 130° - \gamma' \approx 130° - 113°10' = 16°50'$, and

$$a' = \frac{10 \sin \alpha'}{\sin 50°} \approx \frac{10(0.2896)}{0.7660} \approx 3.78$$

A scale drawing of the triangles is shown in Figure 6–23.

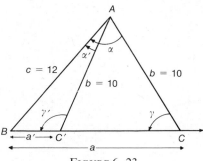

FIGURE 6–23

It may occur to you that in a practical application the solution would not be so ambiguous as presented here, since the problem-solver would more than likely know which triangle he or she is trying to solve. This may not always be the case, however, because if we know that an airplane leaves an airport at point A, flies 100 miles to point B, changes course and flies 50 miles to point C so that the measure of angle CAB is $10°$, we don't know whether the airplane is at point C_1 or C_2 in Figure 6–24. A radarscope located at A would indicate the correct position, of course, and hopefully the pilot would know whether she was flying toward or away from the airport.

You should be aware of the existence of the ambiguous case, however, and should exercise some care when the possibility of two solutions presents itself.

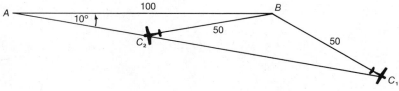

FIGURE 6–24

EXERCISES

Determine the number of possible solutions, then, if possible, solve each of the following triangles.

1. $a = 12, b = 8, \alpha = 36°$
2. $a = 12, b = 8, \alpha = 136°$
3. $b = 150, c = 200, \beta = 20°20'$
4. $a = 4.50, c = 3.25, \gamma = 31°50'$
5. $a = 63, c = 150, \alpha = 24°50'$
6. $b = 200, c = 195, \gamma = 77°10'$

6.3 THE LAW OF COSINES

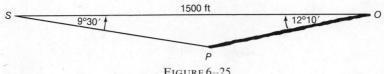

FIGURE 6-25

7. $\alpha = 135°20'$, $a = 35.2$, $b = 3.52$
8. $\alpha = 23°10'$, $\beta = 49°50'$, $c = 212$
9. $b = 303$, $c = 129$, $\gamma = 28°10'$
10. $\alpha = 83°40'$, $\beta = 47°50'$, $\gamma = 48°30'$
11. $a = 512$, $b = 475$, $\beta = 65°30'$
12. $a = 6.50$, $c = 8.08$, $\alpha = 53°50'$

13. An airplane flies 200 miles from A to B, then changes course and flies 75 miles from B to C. If the angle BAC measures $5°$, how far is it from C to A?

14. The earth is approximately 9.29×10^7 miles from the sun and the planet Venus is approximately 6.72×10^7 miles from the sun. If an observer on earth calculates the angle between Venus and the sun, with the earth at the vertex, to be $8°40'$, what would be the distance from the earth to Venus at that time?

15. An outcropping of mineral ore occurs at point O as indicated in Figure 6-25. A shaft is started 1500 feet from O at point S, at an angle of $9°30'$ with the ground. If the angle SOP measures $12°10'$, find the distance from S to P.

16. A sailboat runs 350 yards from P to Q, then changes course and sails 200 yards from Q to R. If angle QPR measures $42°10'$ and $\angle PQR$ is acute, what is the distance from P to R?

17. Can you solve the triangle with sides $a = 30$, $b = 40$, and $c = 45$ by a direct application of the Law of Sines? Demonstrate the reason for your answer.

18. Solve the triangle in problem 17 without using the Law of Cosines, which is introduced in the following section. *Hint*: *Draw the altitude from C to side AB, and use the Pythagorean relation on the two right triangles which are formed.*

6.3 THE LAW OF COSINES

Those who solved problem 18 in the exercises in Section 6.1 by means of right triangles will be pleased to find out about the LAW OF COSINES if they have not used it before. In that problem the triangle to be solved had $\alpha = 40°$, $b = 20$, and $c = 15$, or two sides and the included angle, sometimes known as SAS. In problem 17 in Section 6.1, it was demonstrated that the problem could not be solved directly by the Law of Sines, since each combination of two of the three members of the equation $\dfrac{\sin 40°}{a} = \dfrac{\sin \beta}{20} = \dfrac{\sin \gamma}{15}$ resulted in an equation with two variables.

One way to begin the solution of this problem would be to use the distance formula to find the length of side a. We can place coordinate axes on the triangle as shown in Figure 6–26, and by the distance formula

$$a = \sqrt{(x-20)^2 + (y-0)^2} = \sqrt{x^2 - 40x + 400 + y^2}$$
$$= \sqrt{(x^2 + y^2) + 400 - 40x}$$

We note that $x^2 + y^2 = c^2 = 15^2 = 225$, and since $\cos 40° = x/15$, $x = 15 \cos 40°$. Replacing the variables in the right member, then, $a = \sqrt{225 + 400 - 40(15 \cos 40°)}$, or $a = \sqrt{625 - 600 \cos 40°}$, and if we want a decimal approximation, $a \approx \sqrt{625 - 600(.7660)} = \sqrt{165.4} \approx 12.9$. To complete the problem we could then use the Law of Sines.

The above was fairly simple, but the same method could be applied with other triangles in which two sides and their included angle are given. This usually inspires mathematicians to develop a general formula. This time let α be the measure of an obtuse angle, as indicated in Figure 6–27. Again we use the distance formula to obtain

$$a = \sqrt{(x-b)^2 + (y-0)^2} = \sqrt{x^2 - 2bx + b^2 + y^2}$$
$$= \sqrt{b^2 + (x^2 + y^2) - 2bx}$$

and again we note that $x^2 + y^2 = c^2$, and that since $\cos \alpha = x/c$, $x = c \cos \alpha$.

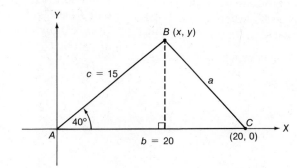

FIGURE 6–26

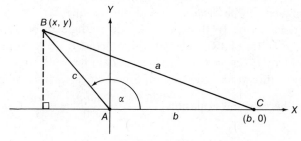

FIGURE 6–27

6.3 THE LAW OF COSINES

Thus we can write

$$a = \sqrt{b^2 + c^2 - 2bc \cos \alpha}$$

and if both sides are squared the result is the LAW OF COSINES,

$$a^2 = b^2 + c^2 - 2bc \cos \alpha \tag{1}$$

This is not quite as easy to recall as the Law of Sines at first, but it helps to say it in words: *The square of the length of one side of a triangle is equal to the sum of the squares of the lengths of the other two sides less twice their product multiplied by the cosine of the angle opposite the first side.* We could just as easily place the coordinate axes at the vertex of each of the other two angles and demonstrate that

$$b^2 = a^2 + c^2 - 2ac \cos \beta \tag{2}$$

$$c^2 = a^2 + b^2 - 2ab \cos \gamma \tag{3}$$

Had we applied the Law of Cosines to our original problem, the one in which we were given $\alpha = 40°, b = 20, c = 15$, the result would have been simply

$$a^2 = 20^2 + 15^2 - 2(20)(15) \cos 40°$$
$$a^2 = 625 - 600 \cos 40°$$

which is quite direct.

EXAMPLE 1 Solve the triangle for which $a = 30, c = 25$, and $\beta = 110°$.

SOLUTION Sketch the triangle and label the given parts, as shown in Figure 6–28. Then $b^2 = a^2 + c^2 - 2ac \cos \beta$, so we have $b^2 = 30^2 + 25^2 - 2(30)(25) \cos 110° = 900 + 625 - 1500 \cos 110°$. Don't forget that since $110°$ is the measure of a second quadrant angle, $\cos 110° = -\cos 70°$, so

$$b^2 = 1525 - 1500(-\cos 70°)$$
$$b^2 = 1525 + 1500 \cos 70°$$
$$b = \sqrt{1525 + 1500 \cos 70°}$$

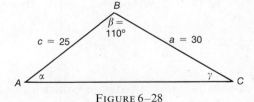

FIGURE 6–28

If we want a numerical approximation, $b \approx \sqrt{1525 + 1500(.3420)} = \sqrt{2038} \approx 45.1$. Then

$$\frac{\sin \alpha}{30} \approx \frac{\sin 110°}{45.1} \approx \frac{\sin \gamma}{25}$$

so
$$\sin \alpha \approx \frac{30 \sin 110°}{45.1} \approx \frac{30(0.9397)}{45.1} \approx 0.6251$$

and $\alpha \approx 38°40'$, while

$$\sin \gamma \approx \frac{25 \sin 110°}{45.1} \approx \frac{25(0.9397)}{45.1} \approx 0.5209$$

and $\gamma \approx 31°20'$.

EXAMPLE 2 Find γ if $a = 12$, $b = 15$, and $c = 20$.

SOLUTION This isn't the case with two sides and their included angle given, but since

$$c^2 = a^2 + b^2 - 2ab \cos \gamma$$

only one variable remains after substituting the given values. Thus we have

$$20^2 = 12^2 + 15^2 - 2(12)(15) \cos \gamma$$
$$400 = 144 + 225 - 360 \cos \gamma$$
$$360 \cos \gamma = 144 + 225 - 400 = -31$$
$$\cos \gamma = \frac{-31}{360}$$

from which $\gamma = \text{Arc} \cos \left(\frac{-31}{360}\right)$.

Since $\cos \gamma$ is negative, γ must be obtuse. To find a numerical approximation, then, we must use

$$\gamma = \text{Arc} \cos \left(\frac{31}{360}\right) \approx \text{Arc} \cos (0.0861) \approx 85°00'$$

and $\gamma \approx 180° - 85°00' = 95°00'$.

The case we have just solved is that which involves three given sides, or SSS, which also could not be solved directly by means of the Law of Sines. We could simplify the work in this case, however, if we were to solve equations (1), (2), and (3) for

6.3 THE LAW OF COSINES 163

cos α, cos β, or cos γ. Solving equation (1), we have

$$a^2 = b^2 + c^2 - 2bc \cos \alpha$$

$$2bc \cos \alpha = b^2 + c^2 - a^2$$

$$\cos \alpha = \frac{b^2 + c^2 - a^2}{2bc} \tag{4}$$

Note that we still have cos α and a on the outside of the equation.

The Law of Cosines has a wide variety of applications, and is often used in conjunction with the Law of Sines. First we need to consider a method of locating objects relative to a given point which is widely used in surveying and the armed forces, called the BEARING of the object from the given point. If the given point is at the origin, then the vertical line through the origin, or the y axis, is considered as pointing north and at 0°, while the positive direction of the horizontal x axis is pointing east and at 90°. Thus the bearing system measures angles clockwise from 0° through 360°. In Figure 6–29 the bearing of P from point 0 is 62°, the bearing of Q from 0 is 110°, of R is 180°, S is 225°, and T is 350°. When there are several points involved and we want to consider their bearings from one another, we can draw a vertical line through each point, as in Figure 6–30. Then if the bearing of B from A is 30°, of C from B is 130°, and of A from C is 260°, we can figure different angles and bearings as follows: $m°(\angle S'BA) = 30°$, since $\angle S'BA$ and $\angle NAB$ are alternate interior angles; $m°(\angle S'BC) = 50°$, since $\angle N'BC$ and $\angle S'BC$ are supplementary and their sum is 180°; $m°(\angle N''CB) = 50°$, since $\angle N'BC$ and $\angle N''CB$ are interior angles on the same side of transversal BC and are thus supplementary; $m°(\angle S''CA) = 80°$, since $m°(\angle N''CS'') = 180°$ and $260° - 180° = 80°$; $m°(\angle BCA) = 50°$, since $m°(\angle BCA) = 180° - m°(\angle N''CB) - m°(\angle S''CA)$; and you should by now be able to determine that $m°(\angle SAC) = 100°$ and $m°(\angle BAC) = 50°$. The bearing of C from A is 80°; the bearing of A from B is

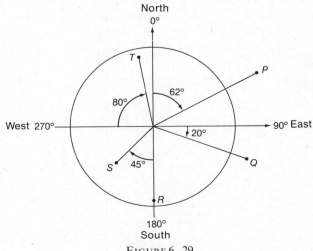

FIGURE 6–29

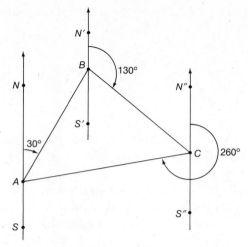

FIGURE 6-30

210°, and the bearing of B from C is 310°. These can be obtained from the given bearings by simply adding or subtracting 180°. There are many alternative methods of obtaining these angle measures, depending upon how you look at them.

EXAMPLE 3   The control tower of an airport observes an executive jet 50 miles away on a bearing of 100° from the airport and a cargoliner 60 miles away on a bearing of 110° from the airport. What is the distance between the two planes at that instant?

SOLUTION   Draw a sketch such as that in Figure 6-31 with the airport at A, the executive jet at B, and the cargoliner at C. Remember that bearings are being given now, and that the angles are measured clockwise with 0° being the vertical ray. It is evident that $m°(\angle BAC) = 110° - 100° = 10°$, and we have two sides and their included angle. Thus if $d = BC$, we have $d^2 = 50^2 + 60^2 - 2(50)(60)\cos 10° = 2500 + 3600 - 6000\cos 10° = 6100 - 6000\cos 10°$, and $d = \sqrt{6100 - 6000\cos 10°}$. A numerical approximation is more than likely required here, so $d \approx \sqrt{6100 - 6000(0.9848)} = \sqrt{191.2} \approx 13.8$.

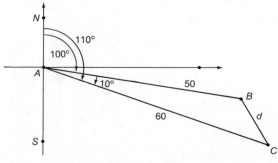

FIGURE 6-31

6.3 THE LAW OF COSINES

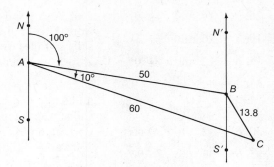

FIGURE 6-32

As a further illustration of the use of the Law of Sines in this particular situation, suppose the cargoliner drops a mysterious package by parachute at the instant mentioned above, and the pilot of the executive jet wants to fly over to see who is waiting for it on the ground. What course must be flown in order to reach point C? This calls for a bit more work and an addition to our previous figure, as indicated in Figure 6-32. Now we must find $m°(\angle N'BC)$, which in this case will represent the course from B to C. We know $m°(\angle NAB) = m°(\angle ABS') = 100°$, since $\angle NAB$ and $\angle ABS'$ are alternate-interior angles and must have the same measure. If we can find $m°(\angle ABC)$, we will then know $m°(\angle CBS') = m°(\angle ABC) - m°(\angle ABS')$, and $m°(\angle N'BC) = 180° - m°(\angle CBS')$. (The last sentence illustrates the distinct advantage of having an instructor who can point things out to the student.)

If $\beta = m°(\angle ABC)$, then

$$\frac{\sin 10°}{13.8} \approx \frac{\sin \beta}{60}$$

$\sin \beta \approx \dfrac{60 \sin 10°}{13.8} \approx \dfrac{60(0.1736)}{13.8} \approx 0.7548$, and $\beta \approx 49°00'$. Thus $m°(\angle N'BC) \approx 180° - 49°00' = 131°00'$ is the course.*

EXERCISES

Use the Law of Cosines to solve for the indicated part of the given triangle in problems 1–13.

1. Find b if $a = 8$, $c = 10$, and $\beta = 30°$
2. Find c if $a = 10$, $b = 12$, and $\gamma = 45°$
3. Find a if $b = 12$, $c = 20$, and $\alpha = 120°$
4. Find b if $a = 15$, $c = 20$, and $\alpha = 150°$
5. Find c if $a = 30$, $b = 25$, and $\gamma = 50°$
6. Find a if $b = 8.0$, $c = 6.5$, and $\alpha = 75°$

---

* It is highly unlikely that either the pilot or the control tower could have the time to solve these problems in this manner. Various types of inflight computers give rapid solutions very easily, but the *basis* for the computer solutions is here, and these methods help in the programming of some of the computers, so it is worthwhile and perhaps interesting to do these this way while we are still on the ground.

7. Find α if $a = 25$, $b = 15$, and $c = 20$
8. Find β if $a = 10$, $b = 26$, and $c = 24$
9. Find γ if $a = 15$, $b = 12$, and $c = 10$
10. Find α if $a = 6.0$, $b = 7.5$, and $c = 9.0$
11. Find β if $a = 3.2$, $b = 5.5$, and $c = 4.0$
12. Find γ if $a = 30$, $b = 42$, and $c = 65$
13. Find c if $a = 25$, $b = 32$, and $\gamma = 90°$
14. Find c if $a = a$, $b = b$, and $\gamma = 90°$. Comment on the result.

15. Solve equation (2) of this section in order to obtain an expression for $\cos \beta$ in terms of a, b, and c.

16. Solve equation (3) of this section in order to obtain an expression for $\cos \gamma$ in terms of a, b, and c.

Solve the following triangles:

17. $a = 12$, $c = 20$, $\beta = 82°30'$
18. $b = 15$, $c = 30$, $\alpha = 20°20'$
19. $a = 2.5$, $b = 3.2$, $\gamma = 144°50'$
20. $a = 4.0$, $c = 4.0$, $\beta = 100°10'$
21. $a = 2.9$, $b = 3.1$, $c = 3.5$
22. $a = 38$, $b = 24$, $c = 32$
23. $a = 110$, $b = 220$, $c = 300$
24. $a = 24$, $b = 45$, $c = 51$

25. A helicopter flies 55 miles from A to B, turns right 35°, then flies 70 miles to C. How far is it from C to A? Draw a sketch and be sure to use the correct angle.

26. A ship sails 20 miles on a course of 220° from X to Y, then changes course to 290° and sails 22 miles from Y to Z. How far is it from Z to X and what is the bearing from X to Z?

27. A fishing trawler leaves port at 5:30 A.M. and sails on a course of 310° at 12 knots. A Coast Guard patrol boat leaves the same port at 6:00 A.M. and sails on a course of 230° at 18 knots. At 8:30 A.M. the trawler sends a distress signal. How far is the patrol boat from the trawler, and what course must it sail in order to reach the trawler if the trawler can no longer move?

28. Find the perimeter of a regular pentagon if it is inscribed in a circle with radius 50 cm.

29. Find the area of a triangle which has sides which measure 30 inches, 35 inches, and 42 inches.

30. Find the area of a triangle which has sides which measure 2.6 m, 3.6 m, and 5.5 m.

31. Whether a triangle is acute, right, or obtuse depends upon the relationship between the longest side and the other two sides. What is that relationship? Use your results to determine whether triangles having sides with the given measures are acute, right, or obtuse.

   a. 8, 9, 10   *b.* 8, 15, 17
   c. 8, 15, 20   *d.* 10, 15, 18
   e. 16, 20, 26   *f.* 28, 98, 100

32. Use the Law of Cosines to find c in the triangle with $a = 30$, $b = 25$, and $\alpha = 60°$. *Hint:* Substitute the given values into equation (1) of this section. A quadratic equation will result.

6.4 VECTORS

Many applications of trigonometry occur in various aspects of work with vectors, which involve quantities having both magnitude and direction. One example of an application of vector quantities relates to the flight of an airplane as affected

6.4 VECTORS

by the wind. It is simple to calculate the groundspeed of an airplane if its airspeed is 150 miles per hour and there is a headwind or tailwind of 20 miles per hour, but what happens if the airplane is on a course of 270° and the wind is blowing from 220°? We can answer the question with a simple application of vectors, as we shall see later.

A GEOMETRIC VECTOR is a directed line segment which has an initial point and a terminal point. In Figure 6–33 the two points A and B determine two different vectors, the directed line segment AB and the directed line segment BA. To indicate the difference we shall use the notation $\overrightarrow{AB}$ when the initial point is A and the terminal point is B, and $\overrightarrow{BA}$ when the initial point is B and the terminal point is A. The length of segment AB is called the MAGNITUDE of the vector, and since $\overrightarrow{AB}$ and $\overrightarrow{BA}$ involve the same segment their magnitudes are equal. We may use the symbol $|\overrightarrow{AB}|$ to represent the magnitude of $\overrightarrow{AB}$, thus $|\overrightarrow{AB}| = |\overrightarrow{BA}|$.

The DIRECTION of a vector is determined with respect to a coordinate system. If we place the origin of the system at the initial point of the vector, then the DIRECTION ANGLE can be specified as the angle formed by the segment and the positive ray of the x axis, as in Figure 6–34a, or between the segment and the 0° or north ray in a bearing system as in Figure 6–34b.

Two vectors are said to be equal if they have the same direction and the same magnitude. In Figure 6–35 the vectors $\overrightarrow{AB}$ and $\overrightarrow{GH}$ are equal, but $\overrightarrow{AB} \neq \overrightarrow{CD}$ since their directions are not equal, and $\overrightarrow{AB} \neq \overrightarrow{EF}$ since $|\overrightarrow{AB}| \neq |\overrightarrow{EF}|$.

In order to add two vectors $\overrightarrow{AB}$ and $\overrightarrow{CD}$ we locate a vector $\overrightarrow{C'D'} = \overrightarrow{CD}$ so that the initial point of $\overrightarrow{C'D'}$ is at the terminal point of $\overrightarrow{AB}$, and $\overrightarrow{AD'}$ is the sum of $\overrightarrow{AB}$ and $\overrightarrow{CD}$, as indicated in Figure 6–36. The vector sum of two or more vectors is called their RESULTANT, and if $\vec{v} = \vec{v}_1 + \vec{v}_2$, it can be shown that $\vec{v}_1 + \vec{v}_2 = \vec{v}_2 + \vec{v}_1$ (Figure 6–37). A complete algebra of vectors can be developed.

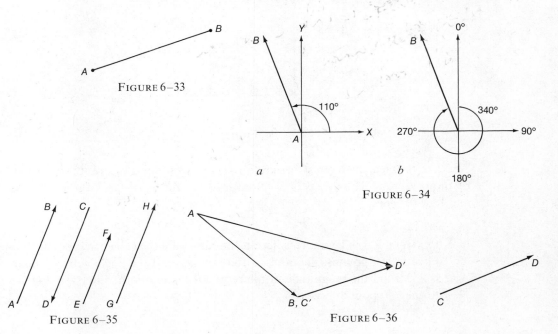

FIGURE 6–33

FIGURE 6–34

FIGURE 6–35

FIGURE 6–36

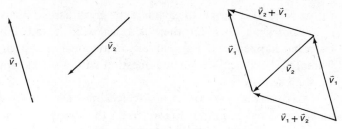

FIGURE 6–37

One problem of interest in physics is that of RESOLVING a vector into its horizontal and vertical COMPONENTS. This means that given a vector $\vec{v}$, we want to find a horizontal vector, $\vec{v}_x$, and a vertical vector $\vec{v}_y$, such that $\vec{v}_x + \vec{v}_y = \vec{v}$.

EXAMPLE 1   Given the vector $\vec{v}$ with $|\vec{v}| = 12$ and direction angle $50°$, find $|\vec{v}_x|$ and $|\vec{v}_y|$.

SOLUTION   Draw the given vector $\vec{v}$ as shown in Figure 6–38. Then $\vec{v}_x$ and $\vec{v}_y$ are indicated on the x and y axes. The solution is a simple application of right triangle trigonometry, that is, $\cos 50° = |\vec{v}_x|/12$, so $|\vec{v}_x| = 12 \cos 50° \approx 12(0.6428) \approx 7.71$, and $\sin 50° = |\vec{v}_y|/12$, so $|\vec{v}_y| = 12 \sin 50° \approx 12(0.7660) \approx 9.19$.

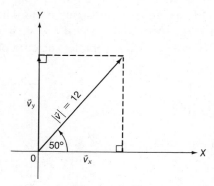

FIGURE 6–38

Another problem of interest in physics is that of finding the resultant of two or more vectors. This is easily accomplished by means of the Law of Cosines and Law of Sines.

EXAMPLE 2   Two forces act on the same object at the same point. One force is 150 pounds, the other is 180 pounds, and the angle between them is $25°$. Find the magnitude of the resultant force and the direction relative to the 180 pound force.

SOLUTION The geometric vectors are drawn in Figure 6–39. Although they are acting at the same point, 0, their resultant force $\overline{OR}$ is the sum of the vectors $\overline{OP}$ and $\overline{OQ}$ as indicated in the figure. Since $|\overline{OP}| = |\overline{O'P'}| = |\overline{QR}|$ and the direction of $\overline{OP}$ is the same as the direction of $\overline{QR}$, we have $m°(\angle OQR) = 180° - 25° = 155°$, and thus $|\overline{OR}|^2 = 180^2 + 150^2 - 2(180)(150) \cos 155°$ by the Law of Cosines. Simplifying we obtain

$$|\overline{OR}|^2 = 32400 + 22500 - 54000 \cos 155°$$
$$= 54900 - 54000(-\cos 25°)$$
$$= 54900 + 54000 \cos 25°$$
$$|\overline{OR}| = \sqrt{54900 + 54000 \cos 25°}$$

A decimal approximation is usually wanted, so

$$|\overline{OR}| \approx \sqrt{54900 + 54000(0.9063)}$$
$$\approx \sqrt{103840} \approx 322.$$

To find the direction of $\overline{OR}$ we use the Law of Sines. If $\alpha = m°(\angle ROQ)$, $\dfrac{\sin \alpha}{150} = \dfrac{\sin 155°}{|\overline{OR}|}$, so $\sin \alpha = \dfrac{150 \sin 155°}{|\overline{OR}|} \approx \dfrac{150 \sin 25°}{322} \approx \dfrac{150(0.4226)}{322} \approx 0.1969$, $\alpha \approx 11°20'$.

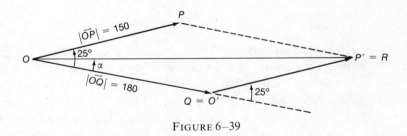

FIGURE 6–39

A final example of an application of vector relationships is that of the airplane mentioned at the beginning of the section. If an airplane starts from point A and heads toward point B, yet there is a wind blowing from the side, then the result will be that the airplane will actually fly to point C. In Figure 6–40 we see that although the airplane is actually headed for B, its track is along the line AC. The situation with respect to vectors is that if there is a wind, the course and groundspeed vector is the resultant of the wind velocity vector added to the heading and airspeed vector. The effects of a headwind or a tailwind are what we would expect, while the effect of a wind from some other angle calls for more calculation. In Figure 6–41, we denote the wind speed vector by $\overline{W}$, the heading and airspeed vector by $\overline{H}_a$, and the course and groundspeed vector by $\overline{C}_g$. The angle ω (Greek

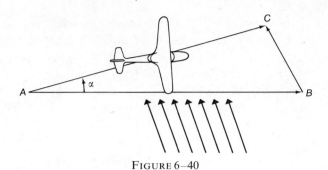

FIGURE 6–40

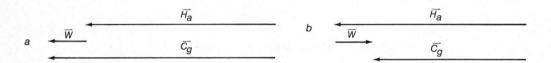

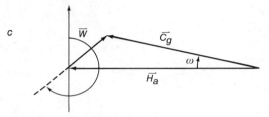

FIGURE 6–41

letter omega) between the heading and course vectors in Figure 6–41c is called the DRIFT ANGLE. We see that a 20 mph tailwind increases the speed of the airplane from 150 to 170 mph, a 20 mph headwind decreases the speed from 150 to 130 mph, and a 20 mph wind from 220° decreases the speed from 150 to about 140 mph. We can calculate the effect more accurately by means of the Law of Cosines. In Figure 6–42 we see that $m°(\angle N'QR)$ must be 40°, and $m°(\angle PQN') = 90°$, so $m°(\angle RQP) = 50°$. Then since $|\vec{H_a}| = 150$ and $|\vec{W}| = 20$, $|\vec{C_g}|^2 = 150^2 + 20^2 - 2(150)(20)\cos 50°$, or, simplifying

$$|C_g|^2 \approx 22500 + 400 - 6000(0.6428)$$

$$\approx 19043$$

$$|\vec{C_g}| \approx 138 \text{ mph}$$

A more practical problem, of course, would be the case in which a pilot wants to complete a trip in a certain time and knows the velocity of the wind at the time of takeoff. The pilot needs to know what heading and airspeed must be maintained in order to be on schedule.

6.4 VECTORS 171

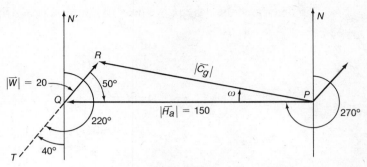

FIGURE 6–42

EXAMPLE 3 A pilot wants to fly from point A to point B, a distance of 300 miles, in $2\frac{1}{2}$ hours. The bearing of B from A is 035°, and the wind is blowing from 345° at 30 mph. What heading and airspeed should be used in order to arrive at the destination on time?

SOLUTION Draw the appropriate vectors as shown in Figure 6–43. Since $d = rt$, $r = d/t$ and the ground speed, then, is $300 \div 2\frac{1}{2} = 120$ mph. Since BC is parallel to AS, $m°(\angle N'BC) = m°(\angle NAS) = 15°$. Since AN is parallel to BN', $m°(\angle NAB) = m°(\angle TBA) = 35°$. Thus $m°(\angle ABC) = 180° - 15° - 35° = 130°$, and $|\overline{W}| = 30$, so

$$|\overline{H}_a|^2 = 120^2 + 30^2 - 2(120)(30) \cos 130°$$
$$\approx 14400 + 900 - 7200(-0.6428)$$
$$\approx 19928$$
$$|\overline{H}_a| \approx 141 \text{ mph}.$$

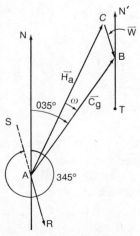

FIGURE 6–43

To determine the heading we need to know $m° (\angle CAB) = \omega$, and

$$\frac{\sin \omega}{30} = \frac{\sin 130°}{|\overline{H}_a|}$$

$$\sin \omega = \frac{30 \sin 130°}{|\overline{H}_a|}$$

$$\approx \frac{30(0.7660)}{141}$$

$$\approx 0.1630$$

$$\omega \approx 9°20'$$

The heading must be $035° - 9°20'$, then, or $025°40'$. Again we concede that in practice other methods would be used. Those of you who are pilots might want to check results with your inflight computers.

EXERCISES

Determine the magnitudes of the horizontal and vertical components of the vectors whose magnitudes and direction angles are given in problems 1–4.

1. 10, 75°
2. 15, 23°41'
3. 5.6, 145°
4. 8.0, −42°20'

A vector could be described in polar coordinate form, so that $\bar{v} = (5, 140°)$ would be a vector with magnitude 5 and direction angle 140°. Find the resultant of the pairs of vectors given in problems 5–10.

5. $\bar{v}_1 = (7, 0°), \bar{v}_2 = (4, 90°)$
6. $\bar{v}_1 = (6, 180°), \bar{v}_2 = (8, 90°)$
7. $\bar{v}_1 = (8, 0°), \bar{v}_2 = (10, -90°)$
8. $\bar{v}_1 = (5, 180°), \bar{v}_2 = (6, -90°)$
9. $\bar{v}_1 = (5.6, 20°), \bar{v}_2 = (6.4, -80°)$
10. $\bar{v}_1 = (8.0, -32°), \bar{v}_2 = (9.5, -152°)$

11. Forces of 12 pounds and 15 pounds act on a point with an angle of 20° between them. Find the magnitude of the resultant force.

12. Forces of 110 pounds and 190 pounds act on a point with an angle of 40° between them. Find the magnitude of the resultant force and the angle between it and the 110 pound force.

13. A pilot wants to fly on a course of 320° with a groundspeed of 100 mph. If the wind is from 060° at the rate of 25 mph, what heading and airspeed must be maintained?

14. A pilot wants to fly on a course of 210° with a groundspeed of 550 mph. If the wind is from 350° at the rate of 45 mph, what heading and airspeed must be maintained?

15. A pilot who flies on a heading of 105° at an airspeed of 130 mph finds that he has actually been on a course of 113° and has made good a groundspeed of 145 mph. What were the direction and rate of the wind?

16. A pilot who wants to fly on a course of 283° at a groundspeed of 200 mph finds that she must maintain a heading of 280° to stay on course. If the wind is from 160°, find the rate of the wind.

17. If $\overline{A}_x, \overline{A}_y$ and $\overline{B}_x, \overline{B}_y$ are the horizontal and vertical components of vectors $\overline{A}$ and $\overline{B}$, respectively, it can be shown that $\overline{V} = \overline{A} + \overline{B}$ can be determined by using $\overline{V} = \overline{V}_x + \overline{V}_y$, where $\overline{V}_x = \overline{A}_x + \overline{B}_x$ and $\overline{V}_y = \overline{A}_y + \overline{B}_y$. Draw the vectors $A = (12, 30°)$ and $B =$

(8, 65°) and their resultant $\vec{V} = \vec{A} + \vec{B}$ and verify this graphically.

18. Find the resultant $\vec{V}$ in problem 17 by calculating $\vec{A}_x, \vec{A}_y, \vec{B}_x, \vec{B}_y$, then check the result by means of the Law of Cosines and Law of Sines.

6.5 MISCELLANEOUS APPLICATIONS

Finding applications of trigonometry is easy—all you have to do is go to a library and browse through books in the science, engineering, and technical sections. By now you have enough background in trigonometry to be able to handle a wide variety of problems. Many applications require more mathematics, particularly calculus and differential equations, along with trigonometry for their solution. In this section you will see just a small sample of applications.

From the **machine shop** comes the following: It is necessary to find the depth of cut h (refer to Figure 6–44) required to machine a vee-shaped channel having known width w and sides inclined at known angles α and β respectively with the vertical, and having an arc of known radius r at the bottom of the vee.* A formula for finding the depth of cut is given in the answers for this section, but first see if you can work out the solution on your own.

1.   Find the depth of cut h for a vee-shaped channel if $\alpha = 25°20'$, $\beta = 32°50'$, $w = 11.50$ inches and $r = 1.25$ inches.

2.   Find the depth of cut h for a vee-shaped channel if $\alpha = 18°10'$, $\beta = 27°20'$, $w = 22.40$ cm and $r = 2.70$ cm.

3.   Derive the formula for finding h given in the answer for problem 1.

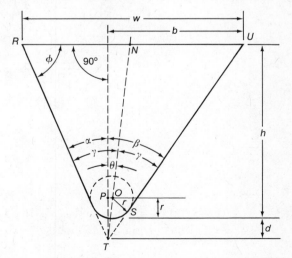

FIGURE 6–44

* This problem and Figure 6–44 are from Holbrook Horton: *Mathematics at Work*, 2nd ed., Industrial Press Inc., 1957, pp. 7–4—7–6. With permission.

Another problem from the machine shop: Find the number of revolutions per minute of a governor (see Figure 6–45) for a specified distance h, if the weight W of the sleeve, the weight L of each ball, the length f of each upper arm, the length g of each lower arm, and the distance s of the lower pivot points from the governor axis are known. If the given dimensions are in inches, the weights are in pounds, and N is the number of revolutions per minute, then

$$N = 188 \sqrt{\frac{\frac{W}{2}\left(1 + \frac{\tan \beta}{\tan \alpha}\right) + L}{Lh}}$$

where $\alpha = \text{Arc sin } (r/f)$ and $\beta = \text{Arc sin } [(r - s)/g]$.*

You will not be asked to derive this formula, but you might want to look it up in *Mathematics at Work*, second edition, by Holbrook L. Horton, published by the Industrial Press.

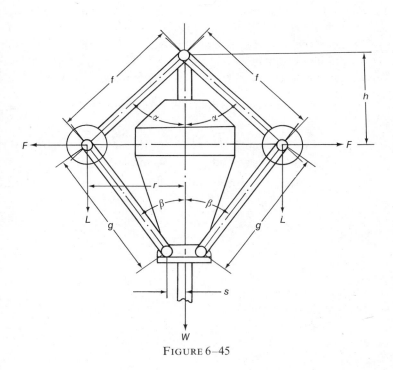

FIGURE 6–45

4 Find N if $W = 100$ pounds, $L = 16$ pounds, $f = 10$ inches, $g = 10$ inches, $s = 1$ inch, and $h = 7.5$ inches.

---

*This formula and Figure 6–45 are from Holbrook Horton: *Mathematics at Work*, 2nd ed., Industrial Press Inc., 1957, pp. 18–22—18–25. With permission.

5 Find N if $W = 120$ pounds, $L = 20$ pounds, $f = 12$ inches, $g = 12$ inches, $s = 1.5$ inches, $h = 8.0$ inches, and $r = 8.94$ inches.

From the field of **medicine** comes the following: Following an osteotomy (dividing or cutting a piece from a bone) of the tibia (long bone between the knee and the ankle), the effective length of the limb can be calculated by means of the Law of Cosines. In Figure 6–46, AC represents the original length of the tibia, β is the angle of deformity, O is the point at which a cut was made, AO and OD are the osteotomized fragments ($AO + OD = AC$), and AD is the effective length of the limb.

If $\ell = AC$, $d = AO$, and $a = OD$, then it can be shown that, for a fixed angle β, as d increases the effective length of the limb, $AD = \ell'$, decreases until $d = a$, whereupon it begins to increase again as d becomes progressively greater than a. The formula for ℓ' is

$$\ell' = \sqrt{a^2 + d^2 + 2ad \cos \beta}$$

6 Use the Law of Cosines to derive the above formula.

7 If $d = .2\ell$, $a = .8\ell$, and $\beta = 20°$
$\ell' = \sqrt{(.8\ell)^2 + (.2\ell)^2 + 2(.8\ell)(.2\ell) \cos 20°} = .9903\,\ell$. Find ℓ' in terms of ℓ when $\beta = 20°$ and   a. $d = .4\ell$,   b. $d = .5\ell$,   c. $d = .6\ell$,   d. $d = .8\ell$.

8 If $\beta = 35°$, find ℓ' in terms of ℓ when   a. $d = .2\ell$,   b. $d = .3\ell$,   c. $d = .4\ell$,   d. $d = .5\ell$.

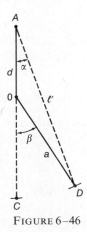

FIGURE 6–46

From **physics**: If a projectile is thrown into the air with an initial velocity v_o at any angle θ_o with the horizontal, then the x and y coordinates of its position at any time t after being projected are given by $x = (v_o \cos \theta_o)t$ and $y = (v_o \sin \theta_o)t - \frac{1}{2}gt^2$, where g is the gravitational constant, or downward acceleration, which is approximately 32 ft/sec². The coordinates, incidentally, are derived from the horizontal and vertical components of the vector $\bar{v}$, which are given by $|\bar{v}_x| = |\bar{v}_o| \cos \theta_o$ and $|\bar{v}_y| = |\bar{v}_o| \sin \theta_o - gt$. Refer to Figure 6–47.

Since t is the same for both the x and y components, $t = x/(v_o \cos \theta_o)$ and y can be expressed in terms of v_o, θ_o, and g, which are constants, as a function of x.

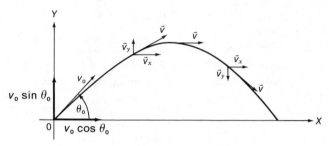

FIGURE 6–47

9 Express y as a function of x. What kind of algebraic equation is this and what is the name of its graph?

10 Plot the graph of the path of a projectile which is fired from the ground at an angle of 45° with the horizontal if its initial velocity is 8 ft/sec.

11 Determine when the projectile in problem 10 reaches   a. its highest point,   b. the ground.

12 What is the highest point reached by the projectile in problem 10?

13 Find the time t_{max} when a projectile fired at an angle of 25° with the horizontal at an initial velocity of 40 ft/sec reaches the highest point in its path. *Hint: At this time the vertical component of velocity, v_y, is equal to zero.*

14 How high does the projectile in problem 13 go?

15 Over what horizontal distance did the projectile in problem 13 travel, and how long was it in the air? *Hint: Find t when $y = 0$.*

A CYCLOID is a curve traced by a point on the circumference of a circle that rolls along a fixed straight line without slipping. If the radius of the circle is a, the x and y coordinates of any point P on the cycloid are given by

$$x = a(\phi - \sin \phi) \text{ and } y = a(1 - \cos \phi)$$

where ϕ is the angle through which the circle has turned (refer to Figure 6–48).

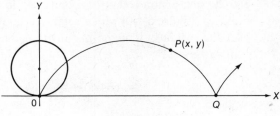

FIGURE 6-48

If the curve OPQ is turned over or reflected with respect to the x axis (refer to Figure 6-49), it then becomes what is known as a BRACHISTOCHRONE or curve of quickest descent, which has the remarkable property that if an object starts from rest at a point O and proceeds to point P' under the influence of gravity alone, the route along the brachistochrone is the path which it will follow in the least amount of time. This fact was proved by the Swiss mathematician John Bernoulli.

16   Derive the equations for the x and y coordinates of the cycloid given above.

17   Plot the graph of a cycloid over $0 \le \phi \le 2\pi$ if $a = 5$. Use an interval of $\phi = \pi/6$.

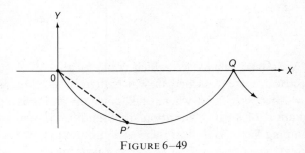

FIGURE 6-49

The ANGULAR VELOCITY of a point moving along the arc of a circle at a constant speed can be found by means of the formula

$$\omega = \frac{\phi}{t}$$

where ω is the angular velocity, ϕ is the measure of the angle of rotation, and t is the time. For example, if a wheel is turning at 30 revolutions per minute, $\phi = 2\pi \cdot 30 = 60\pi$ and $t = 1$ minute, so $\omega = 60\pi/\text{min}$. If another wheel turns through an arc of 10 radians in 2 minutes, its angular momentum is $\omega = 10/(2 \text{ min}) = 5$ radians per minute. Note that ϕ can be expressed as a function of t, that is, $\phi = \omega t$.

18 If a 20 inch piston rod *RP* is connected to the rim of a wheel of radius 5 inches (Figure 6–50) which rotates at 20 revolutions per second, show that the distance x of the piston from the center of the wheel at any time t is given by

$$x = 5 \cos 40\pi t + \sqrt{400 - 25 \sin^2 40\pi t} \quad *$$

19 Find the value of x in the formula of problem 18 at the following values of t: a. $t = 0$ b. $t = 1/80$ sec   c. $t = 1/160$ sec   d. $t = 2.2$ sec

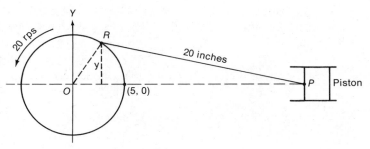

FIGURE 6–50

A DAMPED VIBRATION, which could be that of a spring or of an alternating electrical current, is one in which there is friction or resistance. The formula $y = Ae^{-kt} \cos 2\pi bt$, where A is constant and e is approximately 2.718, is an equation of such a vibration. In Figure 6–51 one such curve is plotted. The damping factor is e^{-kt}. If there were no friction, or no resistance, then the curve would be simply $y = A \cos 2\pi bt$, or a cosine curve with amplitude A. The damping factor causes the amplitude to diminish as time increases, so that it approaches zero as a limit.

20 Find the values of t at which the maximum and minimum values and zeros of the function in the paragraph above occur.

21 Draw the graph of $y = Ae^{(-1/2)t} \cos \pi t$ over the interval $0 \leq t \leq 4$.

---

* The formula and Figure 6–50 are from Frank L. Juszlie, Charles A. Rodgers, *Elementary Technical Mathematics*, 2nd ed., © 1969. Reprinted by permission of Prentice-Hall, Inc., Englewood Cliffs, N.J.

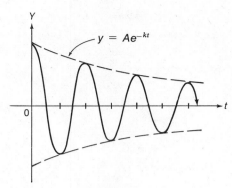

FIGURE 6–51

REVIEW EXERCISES

Solve each of the following triangles:

1. $b = 5, c = 4, \cos \alpha = \frac{1}{8}$.
2. $a = 6, b = 10, \beta = 40°$
3. $b = 8, c = 10, \beta = 67°$
4. $a = 35, b = 25, c = 30$
5. $a = 2.2, c = 3.5, \beta = 125°$
6. $a = 49.2, b = 68.1, \alpha = 36°50'$

Find the area of each of the following triangles:

7. $a = 12, b = 15, \gamma = 52°$
8. $a = 35, b = 25, c = 30$
9. $a = 8, \alpha = 50°, \beta = 60°$

10. Show that the area of a triangle can be found by means of the formula

$$\mathcal{A} = \frac{a^2 \sin \beta \sin (\alpha + \beta)}{2 \sin \alpha}$$

if a, α, and β are known. *Hint:* $\sin(180° - x) = \sin x$.

11. A vector has magnitude 20 and direction angle 48°. Determine the magnitude of its horizontal and vertical components.

12. The sides of a triangle are 8, 10, and 15. Determine the measure of the largest angle.

13. An airplane is flying on a course of 300° at a groundspeed of 100 mph. The wind is blowing from 350° at 20 mph. Determine the airspeed, drift angle, and heading.

14. If the muzzle velocity of a rifle is 2860 ft/sec, and the rifle is fired at an angle of 20° with the horizontal over level ground, find *a.* the time at which the highest point is reached,   *b.* the maximum height of the bullet,   *c.* the horizontal distance traveled,   *d.* the time the bullet is in the air. Refer to the discussion preceding problems 9–15 in Section 6–5, pages 175, 176.

15. Forces of 12 kilograms and 20 kilograms act on an object with an angle of 25° between them. Find the magnitude of the resultant force and the angle it makes with the 20 kg force.

16. A helicopter starts from point A and flies on a course of 220° at 120 mph for 30 minutes to point B, then changes course to 290° and flies at 150 mph for 20 minutes to point C. Find   *a.* the distance from C to A,   *b.* the bearing of A from C, and   *c.* the time required to fly from C to A at 100 mph.

APPENDIX A

Interpolation

At times you may need to calculate values of the trigonometric ratios to the nearest minute, rather than to the nearest ten minutes. The easiest way to do this would be to obtain more extensive tables, such as those appearing in various handbooks of mathematical tables. If you can find the angle to the nearest minute there, however, you may then want the answer to the nearest second. So knowing the method of INTERPOLATION may still prove to be useful.

The graphs of the trigonometric functions are either ascending or descending over the interval from $0°$ through $90°$, or from 0 to $\pi/2$ radians. Suppose that we want to find the value of $\sin 35°14'$. If we greatly enlarge the portion of this graph near $35°$, it will look something like Figure A–1. The section of the curve between A and B is very nearly a straight line. If we enlarge the graph again and exaggerate the curvature we have something which appears like Figure A–2. If points B and D are connected by a straight line, we want to find the length of PT, which is an approximation of the ordinate of point E on the graph and the value of $\sin 35°14'$.

If DQ is parallel to the horizontal axis and DS, PT, and BQ are parallel to the vertical axis, then triangles PDU and BDQ are similar and their corresponding sides are proportional, thus

$$\frac{PU}{BQ} = \frac{DU}{DQ}$$

From the figure it can be seen that $DS = UT = QR = 0.5760$, $DU = 4'$, $DQ = 10'$, and $BQ = BR - QR = 0.5783 - 0.5760 = 0.0023$. Substituting into the proportion, $PU/0.0023 = 4'/10'$, and multiplying both sides of the equation by 0.0023, we find $PU = \dfrac{4'(0.0023)}{10'} = 0.00092 \approx 0.0009$, and $PT = PU + UT = 0.0009 + 0.5760 = 0.5769$. This is exactly the value obtained with a table which

INTERPOLATION

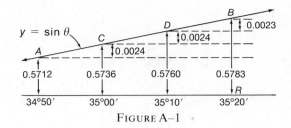

FIGURE A–1

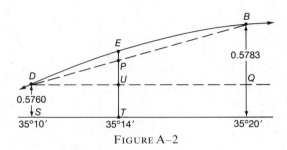

FIGURE A–2

gives the values to the nearest minute. Note that the minute units canceled one another in the computation.

Having observed the foundation for the method, a simpler format for the computation would be that shown below.

$$
\begin{array}{c|cc}
 & \theta & \sin\theta \\
\hline
 & 35°20' & 0.5783 \\
10'\left[4'\left[\begin{array}{c}35°14'\\35°10'\end{array}\right.\right. & & \left.\begin{array}{c}?\\0.5760\end{array}\right]c\quad 0.0023
\end{array}
$$

Other interpolations will be described in the following examples.

EXAMPLE 1   Find the value of tan 48°18′ by interpolation.

SOLUTION   Since 48°18′ is between 48°10′ and 48°20′, look up those values in the table and set up as indicated.

$$
\begin{array}{c|cc}
 & \theta & \tan\theta \\
\hline
 & 48°20' & 1.124 \\
10'\left[8'\left[\begin{array}{c}48°18'\\48°10'\end{array}\right.\right. & & \left.\begin{array}{c}?\\1.117\end{array}\right]c\quad 0.007
\end{array}
$$

Write the proportion

$$\frac{c}{0.007} = \frac{8'}{10'}$$

$$c = \frac{8}{10}(0.007) = 0.0056 \approx 0.006$$

Thus tan 48°18′ ≈ 1.117 + 0.006 = 1.123. Note that we round off c to the nearest thousandth. We do not use more decimal places than in the values from the table.

EXAMPLE 2   Find the value of cos 12°13′ by interpolation.

SOLUTION   Look up the values of cos 12°10′ and 12°20′ and set up as indicated.

$$
\begin{array}{c|cc}
 & \theta & \cos \theta \\ \hline
 & 12°20' & 0.9769 \\
 & 12°13' & ? \\
 & 12°10' & 0.9775 \\
\end{array}
$$

10′ ⎡ 3′ ⎡ 12°13′ ? ⎤ c ⎤ −0.0006
    ⎣   ⎣ 12°10′ 0.9775 ⎦   ⎦

There is a slight complication with cosine, cotangent, and cosecant since their values are decreasing in the interval from 0° through 90°. This is the reason for the difference −0.0006. Writing the proportion

$$\frac{c}{-0.0006} = \frac{3'}{10'}$$

$$c = \frac{3}{10}(-0.0006) = -0.00018 \approx -0.0002$$

Thus cos 12°13′ ≈ 0.9775 − 0.0002 = 0.9773.

The procedure for finding the angle, given the ratio or function value, is nearly the same.

EXAMPLE 3   Find the angle θ for which sin θ = 0.6213, correct to the nearest minute.

SOLUTION   Locate the angles for which the sine values are on either side of 0.6213 and set up as indicated.

INTERPOLATION

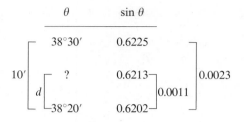

Write the proportion

$$\frac{d}{10'} = \frac{0.0011}{0.0023}$$

$$d = \left(\frac{0.0011}{0.0023}\right)10' \approx (0.4783)(10') = 4.783' \approx 5'$$

Thus $\theta \approx 38°20' + 0°5' = 38°25'$.

EXAMPLE 4 Find the angle θ for which $\cot \theta = 0.4600$, correct to the nearest minute.

SOLUTION Locate the angles for which the cotangent values are on either side of 0.4600.

	θ	$\cot \theta$
	65°20'	0.4592
	?	0.4600
	65°10'	0.4628

Again take note of the decreasing values of the cotangent, but it turns out that d is positive anyway.

$$\frac{d}{10'} = \frac{-0.0028}{-0.0036}$$

$$d = \left(\frac{-0.0028}{-0.0036}\right)10' \approx (0.7778)10' = 7.778' \approx 8'$$

Thus $\theta \approx 65°10' + 0°8' = 65°18'$.

If radian measure is used, we simply turn to Table II in Appendix D and proceed as above.

EXAMPLE 5   Find the value of sin 1.236.

SOLUTION   Look up the values for sin 1.230 = sin 1.23 and sin 1.240 = sin 1.24 and set up as before.

$$\begin{array}{c|cc}
 & \theta & \sin\theta \\
\hline
 & 1.240 & 0.9458 \\
0.010\left[0.006\left[\begin{array}{c}1.236\\ 1.230\end{array}\right.\right. & & \left.\begin{array}{c}?\\ 0.9425\end{array}\right]c\right]0.0033
\end{array}$$

Write the proportion

$$\frac{c}{0.0033} = \frac{0.006}{0.010}$$

$$c = \left(\frac{0.006}{0.010}\right)(0.0033) = 0.00198 \approx 0.0020$$

Thus sin 1.236 ≈ 0.9425 + 0.0020 = 0.9445.

EXAMPLE 6   Find the angle θ, to the nearest thousandth radian, if $\cos\theta = 0.6335$.

SOLUTION   Locate the numbers on either side of 0.6333 and set up as indicated.

$$\begin{array}{c|cc}
 & \theta & \cos\theta \\
\hline
 & 0.890 & 0.6294 \\
 & ? & 0.6335 \\
 & 0.880 & 0.6372
\end{array}$$

Writing the proportion,

$$\frac{d}{0.010} = \frac{-0.0037}{-0.0078}$$

$$d = \left(\frac{-0.0037}{-0.0078}\right)(0.010) \approx (0.4744)(0.010) = 0.004744 \approx 0.005$$

Thus $\theta \approx 0.880 + 0.005 = 0.885$.

EXERCISES

1. Find the value of each of the following by interpolation:
   a. sin 22°19'
   b. cos 40°22'
   c. tan 31°44'
   d. cot 29°12'
   e. tan 52°18'
   f. cos 60°15'
   g. cot 83°29'
   h. sin 48°57'

2. Find the value of each of the following by interpolation:
   a. sin 39°14'
   b. cos 8°48'
   c. tan 15°23'
   d. cot 42°46'
   e. cos 58°11'
   f. sin 76°33'
   g. cot 46°18'
   h. tan 88°47'

3. Find the measure of θ to the nearest minute, if
   a. $\sin \theta = 0.4960$
   b. $\cos \theta = 0.7378$
   c. $\tan \theta = 0.8200$
   d. $\cot \theta = 31.30$
   e. $\tan \theta = 1.703$
   f. $\cos \theta = 0.4000$
   g. $\cot \theta = 0.9100$
   h. $\sin \theta = 0.9623$

4. Find the measure of θ to the nearest minute, if
   a. $\sin \theta = 0.7220$
   b. $\cos \theta = 0.9643$
   c. $\tan \theta = 0.8550$
   d. $\cot \theta = 1.140$
   e. $\cos \theta = 0.5280$
   f. $\sin \theta = 0.3492$
   g. $\cot \theta = 0.0826$
   h. $\tan \theta = 1.261$

5. Find the value of each of the following by means of Table II in Appendix D and interpolation:
   a. sin 0.554
   b. cos 1.036
   c. tan 0.017
   d. cot 0.999

6. Find the value of each of the following by means of Table II in Appendix D and interpolation:
   a. sin 1.358
   b. cos 0.662
   c. tan 1.033
   d. cot 1.521

7. Find the angle θ, in radians, correct to the nearest thousandth, if
   a. $\sin \theta = 0.8000$
   b. $\cos \theta = 0.9935$
   c. $\tan \theta = 2.300$
   d. $\cot \theta = 5.405$

8. Find the angle θ, in radians, correct to the nearest thousandth, if
   a. $\sin \theta = 0.2834$
   b. $\cos \theta = 0.2834$
   c. $\tan \theta = 0.9538$
   d. $\cot \theta = 1.755$

APPENDIX B

Computing Values of Trigonometric Functions

One formula used for computing the value of sin x is

$$\sin x = x - \frac{x^3}{3!} + \frac{x^5}{5!} - \frac{x^7}{7!} + \cdots$$

where $n! = 1 \cdot 2 \cdot 3 \cdots (n-1)(n)$, and x is given in radians. This means that in order to find the value of sin 5°, you must change 5° to radians (5° $= 5 \cdot \pi/180 \approx$ .0873, or simply use Table I in Appendix D) and then

$$\sin 0.0873 \approx 0.0873 - \frac{(0.0873)^3}{1 \cdot 2 \cdot 3} + \frac{(0.0873)^5}{1 \cdot 2 \cdot 3 \cdot 4 \cdot 5} - \frac{(0.0873)^7}{1 \cdot 2 \cdot 3 \cdot 4 \cdot 5 \cdot 6 \cdot 7} + \cdots$$

$$\approx 0.0873 - 0.00011 + 0.0000000423 - 0.00000000000767$$

$$\approx 0.0873 - 0.0001 = 0.0872$$

correct to 4 decimal places. Since the value of the 4th and following terms is so small for angles of 10° or less we can use the first three terms in order to obtain a good approximation to 4 decimal places.

An approximation for cos x can be found by means of the formula

$$\cos x = 1 - \frac{x^2}{2!} + \frac{x^4}{4!} - \frac{x^6}{6!} + \cdots$$

or again only the first three terms in order to find the value to 4 decimal places.

COMPUTING VALUES OF TRIGONOMETRIC FUNCTIONS

EXAMPLE 1   Find the value of cos 6°20′ correct to 4 decimal places.

SOLUTION   6°20′ ≈ 0.1105, so

$$\cos 6°20' \approx \cos 0.1105 \approx 1.0000 - \frac{(.1105)^2}{1 \cdot 2} + \frac{(.1105)^4}{1 \cdot 2 \cdot 3 \cdot 4}$$

$$\approx 1.0000 - 0.0061 + 0.0000$$

$$= 0.9939$$

EXERCISES

Find the value of each of the following correct to four decimal places by means of the formulas in this section.

1. sin 10°
2. sin 8°30′
3. cos 2°40′
4. cos 7°20′
5. sin 0.112
6. cos 0.095

APPENDIX C

Logarithms of the Trigonometric Functions

Some of the calculations required for solution of triangles can be rather tedious without a calculator, so here you will find a brief review of base 10 logarithms, then an introduction to computations using logarithms of the trigonometric functions.

In Figure C–1 the graphs of the function $y = 10^x$ and its inverse, $x = 10^y$, are shown. The inverse, $x = 10^y$, is also a function and we can define $y = \log_{10} x$, or y equals the logarithm of x to the base 10, as an equivalent form to express y in terms of x.

Using the relationship between $x = 10^y$, the exponential form, and $y = \log_{10} x$, the logarithmic form, we find that since

$$10^{-1} = \frac{1}{10}, \quad \log_{10} \frac{1}{10} = -1$$

$$10^0 = 1, \quad \log_{10} 1 = 0$$

$$10^1 = 10, \quad \log_{10} 10 = 1$$

$$10^2 = 100, \quad \log_{10} 100 = 2$$

$$10^3 = 1000, \quad \log_{10} 1000 = 3$$

It can also be proved that if M and N are positive numbers,

$$\log_{10} (M \cdot N) = \log_{10} M + \log_{10} N, \tag{1}$$

$$\log_{10} \left(\frac{M}{N}\right) = \log_{10} M - \log_{10} N \tag{2}$$

$$\log_{10} M^r = r \log_{10} M \tag{3}$$

LOGARITHMS OF THE TRIGONOMETRIC FUNCTIONS

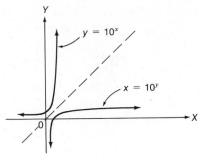

FIGURE C–1

Proof of (1): Let $a = \log_{10} M$ and $b = \log_{10} N$; then $10^a = M$ and $10^b = N$ by definition. Since

$$M \cdot N = 10^a \cdot 10^b$$
$$M \cdot N = 10^{a+b},$$

and
$$\log_{10} (M \cdot N) = a + b$$

again by the definition. Replacing a and b with their equivalents,

$$\log_{10} (M \cdot N) = \log_{10} M + \log_{10} N.$$

Equations (2) and (3) can be proved in a similar fashion.

Since $\log_{10} 10 = 1$, we can also use equation (3) to develop the special case

$$\log_{10} 10^k = k \tag{4}$$

since $\log_{10} 10^k = k \log_{10} 10 = k \cdot 1 = k$.

From its graph in Figure C–1, we see that $y = \log_{10} x$ is an increasing function, and if $1 \leq a \leq 10$, then $\log_{10} 1 \leq \log_{10} a \leq \log_{10} 10$ or $0 \leq \log_{10} a \leq 1$. Values of $\log_{10} a$ for a between 1 and 10 have been calculated and are tabulated in Table III in Appendix D. To find the logarithm of 2.34, for example, look in the left hand column of the table for 2.3, then go across to the right until you reach the column headed by the number 4. You will find $\log_{10} 2.34 = 0.3692$. Similarly $\log_{10} 5.67 = 0.7536$ and $\log_{10} 8.02 = 0.9042$.

In order to find the logarithms of numbers that are not between 1 and 10, we can make use of equations (1) and (4) above. For example, $\log_{10} 234 = \log_{10} (2.34 \times 10^2) = \log_{10} 2.34 + \log_{10} 10^2 = \log_{10} 2.34 + 2 = 0.3692 + 2 = 2.3692$. To find $\log_{10} 56{,}700$, write $\log_{10} 5.67 \times 10^4 = \log_{10} 5.67 + \log_{10} 10^4 = \log_{10} 5.67 + 4 = 0.7536 + 4 = 4.7536$. The base 10 logarithm of 0.00802 is $\log_{10} 8.02 \times 10^{-3} = \log_{10} 8.02 + \log_{10} 10^{-3} = 0.9042 + (-3)$.

You can see that each of the above logarithms involves two numbers—the decimal which is found in the table, and the integer determined by the exponent on 10 when the number is written in the form $a \times 10^k$. The decimal part is called the MANTISSA and the integer part is called the CHARACTERISTIC of the logarithm. In $\log_{10} 234 = 2.3692$, the mantissa is 0.3692 and the characteristic is 2.

If you are familiar with scientific notation, the procedure for finding the base 10 logarithm is simple—the characteristic is merely the exponent on 10 and the mantissa is found in the table. To write a number in SCIENTIFIC NOTATION, we write it as a number between 1 and 10 multiplied by an integral power of 10, that is,

$$9360 = 9.360 \times 10^3$$
$$204{,}000 = 2.04 \times 10^5$$
$$0.00059 = 5.9 \times 10^{-4}$$
$$0.0681 = 6.81 \times 10^{-2}$$

Locate the decimal point to the right of the first nonzero digit in the number, then count the number of places to the original decimal point. If you count to the right the exponent is positive; if you count to the left the exponent is negative.

Since the mantissas given in our table are all positive, we take pains to keep them that way. When we found $\log_{10} 0.00802 = 0.9042 + (-3)$ above it was left that way, rather than changing to -2.0958, which is actually equal to $(-2) + (-0.0958)$. Rather than write $0.9042 + (-3)$ or $(-3) + (0.0942)$, however, it is customary to add $[10 + (-10)]$ to the number and write $10 + (-3) + (0.9042) + (-10) = 7.9042 - 10$ as the value of $\log_{10} 0.00802$. You can then easily see that $7 - 10 = -3$ is the characteristic. Other examples similar to this are

$$\log_{10} 0.298 = 9.4742 - 10$$
$$\log_{10} 0.000425 = 6.6284 - 10$$
$$\log_{10} 0.00000071 = 3.8513 - 10$$

Computation with logarithms requires further use of the tables. For example, to find the product $P = (835)(0.592)$, we note that

$$\log_{10} P = \log_{10} (835)(0.592)$$
$$= \log_{10} 835 + \log_{10} 0.592$$
$$= 2.9217 + (9.7723 - 10)$$
$$= 12.6940 - 10$$
$$= 2.6940$$

Since $\log_{10} P = 2.6940$, we must find the number for which the logarithm has characteristic 2 and mantissa 0.6940. Looking through Table III in Appendix D, we find $\log_{10} 4.94 = 0.6937$ and $\log_{10} 4.95 = 0.6946$. Since 0.6940 is closer to the first of these, we can say that $P \approx 4.94 \times 10^2 = 494$.

Finding the number a when its logarithm is known is called finding the ANTILOGARITHM of a. If $\log_{10} a = b$, $a = \text{antilog}_{10} b$. Other examples of finding antilogarithms are

$$\text{antilog}_{10} 1.3823 = 2.41 \times 10^1 = 24.1$$
$$\text{antilog}_{10} 3.0148 = 1.03 \times 10^3 = 1030$$
$$\text{antilog}_{10} 8.5864 - 10 = 3.86 \times 10^{-2} = 0.0386$$

LOGARITHMS OF THE TRIGONOMETRIC FUNCTIONS

In each case we have used the number whose logarithm is closest to the given mantissa.

In order to find the logarithm of a trigonometric ratio you could use Table I and then Table III in Appendix D. For example, sin 43° = 0.6820, so $\log_{10}$ (sin 43°) = $\log_{10}$ 0.6820 = 9.8338 − 10. In Table IV in Appendix D, the values have been given directly, so you can simply look them up there. In order to save space the −10 part of the characteristic has been omitted from each logarithm, so this needs to be considered. From Table IV in Appendix D we find

$$\log_{10} \cos 29°50' = 9.9383 - 10$$
$$\log_{10} \tan 48°10' = 10.0481 - 10 = 0.0481$$
$$\log_{10} \sin 86°30' = 9.9992 - 10$$

EXAMPLE 1   Find the value of $a = 5.26 \sin 43°40'$ by means of logarithms.

SOLUTION
$$\log_{10} a = \log_{10} 5.26 + \log_{10} \sin 43°40'$$
$$= 0.7210 + 9.8391 - 10$$
$$= 10.5601 - 10 = 0.5601$$
$$a = \text{antilog}_{10} 0.5601 = 3.63$$

The computation is perhaps easier to carry out if a format such as the following is used:

$$\begin{array}{rl}
\log_{10} 5.26 & = 0.7210 \\
(+) \log_{10} \sin 43°40' & = 9.8391 - 10 \\
\hline
\log_{10} a & = 10.5601 - 10 = 0.5601 \\
a & = 3.63
\end{array}$$

EXAMPLE 2   Use logarithms to find the value of α if $\sin \alpha = \dfrac{22.5 \sin 30°10'}{19.8}$

SOLUTION
$$\begin{array}{rl}
\log_{10} 22.5 & = 1.3522 \\
(+) \log_{10} \sin 30°10' & = 9.7012 - 10 \\
\hline
\log \text{ numerator} & = 11.0534 - 10 \\
(-) \log 19.8 & = 1.2967 \\
\hline
\log_{10} \sin \alpha & = 9.7567 - 10 \\
\alpha & = 34°50'
\end{array}$$

EXAMPLE 3 Use logarithms to find $b = \sqrt{45.2}$.

SOLUTION $\log_{10} b = \log_{10} \sqrt{45.2} = \log_{10} (45.2)^{1/2}$
$= \tfrac{1}{2} \log_{10} 45.2$
$= \tfrac{1}{2}(1.6551)$
$= 0.8276$

Thus $b = \text{antilog}_{10}\, 0.8276 = 6.72$.

EXERCISES

1. Find the logarithm of each of the following numbers:
   a. 5.37
   b. 3.3
   c. 8
   d. 609
   e. 15,700
   f. 0.52
   g. 0.0003
   h. 0.00295

2. Find the logarithm of each of the following numbers:
   a. 2.99
   b. 6
   c. 5.8
   d. 888
   e. 314,000
   f. 0.782
   g. 0.0057
   h. 0.000006

3. Find the antilogarithm of each of the following numbers, correct to three significant digits:
   a. 0.5988
   b. 0.8802
   c. 2.2648
   d. 3.7612
   e. 8.5933 − 10
   f. 1.9476
   g. 9.2220 − 10
   h. 7.7727 − 10

4. Find the antilogarithm of each of the following numbers, correct to three significant digits:
   a. 0.2122
   b. 0.9335
   c. 1.2330
   d. 4.9380
   e. 9.9354 − 10
   f. 2.8158
   g. 6.5140
   h. 8.3988 − 10

5. Find the value of each of the following:
   a. $\log_{10} \sin 38°50'$
   b. $\log_{10} \tan 62°20'$
   c. $\log_{10} \cos 50°30'$
   d. $\log_{10} \cot 23°40'$

6. Find the value of each of the following:
   a. $\log_{10} \cos 31°10'$
   b. $\log_{10} \tan 88°50'$
   c. $\log_{10} \cot 65°00'$
   d. $\log_{10} \sin 152°40'$

7. Find the acute angle α (nearest 10 minutes) such that
   a. $\log_{10} \sin \alpha = 9.5978 - 10$
   b. $\log_{10} \cos \alpha = 9.8405 - 10$
   c. $\log_{10} \tan \alpha = 0.4760$
   d. $\log_{10} \sin \alpha = 9.9637 - 10$

8. Find the acute angle α (nearest 10 minutes) such that
   a. $\log_{10} \sin \alpha = 9.6692 - 10$
   b. $\log_{10} \cos \alpha = 9.9618 - 10$
   c. $\log_{10} \tan \alpha = 0.3000$
   d. $\log_{10} \sin \alpha = 9.9265 - 10$

9. Use logarithms to calculate the square root of each of the following numbers correct to three significant digits:
   a. 3.87
   b. 58.4
   c. 23,400
   d. 0.0596

10. Use logarithms to calculate the square root of each of the following numbers correct to three significant digits:
    a. 7.85
    b. 9000
    c. 56,900
    d. 0.876

11. Use logarithms to find the value of each of the following, correct to three significant digits:
    a. $h = 42.6 \sin 82°30'$

b. $b = 239 \tan 25°50'$

c. $c = \dfrac{6080}{\cos 43°20'}$

d. $a = \dfrac{1.27 \sin 53°20'}{\sin 23°50'}$

e. $b = \dfrac{33.3 \sin 108°40'}{\sin 60°20'}$

12. Use logarithms to find the value of each of the following, correct to three significant digits:

a. $\mathscr{A} = 23.5(15)(\sin 48°30')$

b. $a = 680 \cot 34°$

c. $c = \dfrac{59.7}{\sin 62°10'}$

d. $a = \dfrac{62.4 \sin 13°30'}{\sin 28°40'}$

e. $b = \dfrac{2800 \sin 143°50'}{\sin 25°20'}$

13. Use logarithms to find the value of an acute angle α (nearest ten minutes), given that

a. $\sin \alpha = \dfrac{35.8}{43.2}$ b. $\tan \alpha = \dfrac{597}{203}$

c. $\cos \alpha = \dfrac{0.333}{0.485}$

d. $\sin \alpha = \dfrac{235 \sin 76°10'}{598}$

14. Use logarithms to find the value of an acute angle β (nearest ten minutes), given that

a. $\cos \beta = \dfrac{4.36}{6.28}$ b. $\cot \beta = \dfrac{48.8}{50.5}$

c. $\sin \beta = \dfrac{566}{666}$

d. $\sin \beta = \dfrac{23.6 \sin 123°10'}{20.9}$

15. Use logarithms to solve each of the following triangles, given that $C = 90°$:

a. $a = 5.09, A = 53°40'$
b. $b = 32.4, A = 29°20'$
c. $b = 980, c = 1320$
d. $a = 0.355, b = 0.458$

16. Use logarithms to solve each of the following triangles, given that $C = 90°$:

a. $a = 22.2, B = 40°20'$
b. $b = 304, B = 82°00'$
c. $a = 0.209, c = 0.555$
d. $a = 833, b = 699$

17. Use logarithms and the Law of Sines to solve each of the following triangles:

a. $\alpha = 29°40', \beta = 103°10', a = 55.9$
b. $a = 506, b = 668, \beta = 59°50'$
c. $a = 356, c = 285, \gamma = 12°10'$

18. Use logarithms and the Law of Sines to solve each of the following triangles:

a. $\alpha = 57°40', \gamma = 63°40', c = 303$
b. $b = 752, c = 893, \gamma = 128°50'$
c. $a = 236, b = 311, \alpha = 33°20'$

APPENDIX D

Tables

I   TRIGONOMETRIC FUNCTIONS, VALUES IN DEGREES
II   TRIGONOMETRIC FUNCTIONS, VALUES IN RADIANS
III   LOGARITHMS, BASE 10
IV   LOGARITHMS OF TRIGONOMETRIC FUNCTIONS
V   SQUARES AND SQUARE ROOTS

TABLE I
TRIGONOMETRIC FUNCTIONS, VALUES IN DEGREES

θ degrees	θ radians	sin θ	csc θ	tan θ	cot θ	sec θ	cos θ		
0° 00′	.0000	.0000	undefined	.0000	undefined	1.000	1.0000	1.5708	90° 00′
10	.0029	.0029	343.8	.0029	343.8	1.000	1.0000	1.5679	50
20	.0058	.0058	171.9	.0058	171.9	1.000	1.0000	1.5650	40
30	.0087	.0087	114.6	.0087	114.6	1.000	1.0000	1.5621	30
40	.0116	.0116	85.95	.0116	85.94	1.000	.9999	1.5592	20
50	.0145	.0145	68.76	.0145	68.75	1.000	.9999	1.5563	10
1° 00′	0.175	.0175	57.30	.0175	57.29	1.000	.9998	1.5533	89° 00′
10	.0204	.0204	49.11	.0204	49.10	1.000	.9998	1.5504	50
20	.0233	.0233	42.98	.0233	42.96	1.000	.9997	1.5475	40
30	.0262	.0262	38.20	.0262	38.19	1.000	.9997	1.5446	30
40	.0291	.0291	34.38	.0291	34.37	1.000	.9996	1.5417	20
50	0.320	.0320	31.26	.0320	31.24	1.001	.9995	1.5388	10
2° 00′	0.349	.0349	28.65	.0349	28.64	1.001	.9994	1.5359	88° 00′
10	.0378	.0378	26.45	.0378	26.43	1.001	.9993	1.5330	50
20	.0407	.0407	24.56	.0407	24.54	1.001	.9992	1.5301	40
30	.0436	.0436	22.93	.0437	22.90	1.001	.9990	1.5272	30
40	.0465	.0465	21.49	.0466	21.47	1.001	.9989	1.5243	20
50	.0495	.0494	20.23	.0495	20.21	1.001	.9988	1.5213	10
3° 00′	.0524	.0523	19.11	.0524	19.08	1.001	.9986	1.5184	87° 00′
10	.0553	.0552	18.10	.0553	18.07	1.002	.9985	1.5155	50
20	.0582	.0581	17.20	.0582	17.17	1.002	.9983	1.5126	40
30	.0611	.0610	16.38	.0612	16.35	1.002	.9981	1.5097	30
40	.0640	.0640	15.64	.0641	15.60	1.002	.9980	1.5068	20
50	.0669	.0669	14.96	.0670	14.92	1.002	.9978	1.5039	10
4° 00′	.0698	.0698	14.34	.0699	14.30	1.002	.9976	1.5010	86° 00′
10	.0727	.0727	13.76	.0729	13.73	1.003	.9974	1.4981	50
20	.0756	.0756	13.23	.0758	13.20	1.003	.9971	1.4952	40
30	.0785	.0785	12.75	.0787	12.71	1.003	.9969	1.4923	30
40	.0814	.0814	12.29	.0816	12.25	1.003	.9967	1.4893	20
50	.0844	.0843	11.87	.0846	11.83	1.004	.9964	1.4864	10
5° 00′	.0873	.0872	11.47	.0875	11.43	1.004	.9962	1.4835	85° 00′
10	.0902	.0901	11.10	.0904	11.06	1.004	.9959	1.4806	50
20	.0931	.0929	10.76	.0934	10.71	1.004	.9957	1.4777	40
30	.0960	.0958	10.43	.0963	10.39	1.005	.9954	1.4748	30
40	.0989	.0987	10.13	.0992	10.08	1.005	.9951	1.4719	20
50	.1018	.1016	9.839	.1022	9.788	1.005	.9948	1.4690	10
6° 00′	.1047	.1045	9.567	.1051	9.514	1.006	.9945	1.4661	84° 00′
		cos θ	sec θ	cot θ	tan θ	csc θ	sin θ	θ radians	θ degrees

TABLE I, CONTINUED
TRIGONOMETRIC FUNCTIONS, VALUES IN DEGREES

θ degrees	θ radians	sin θ	csc θ	tan θ	cot θ	sec θ	cos θ		
6° 00′	.1047	.1045	9.567	.1051	9.514	1.006	.9945	1.4661	84° 00′
10	.1076	.1074	9.309	.1080	9.255	1.006	.9942	1.4632	50
20	.1105	.1103	9.065	.1110	9.010	1.006	.9939	1.4603	40
30	.1134	.1132	8.834	.1139	8.777	1.006	.9936	1.4573	30
40	.1164	.1161	8.614	.1169	8.556	1.007	.9932	1.4544	20
50	.1193	.1190	8.405	.1198	8.345	1.007	.9929	1.4515	10
7° 00′	.1222	.1219	8.206	.1228	8.144	1.008	.9925	1.4486	83° 00′
10	.1251	.1248	8.016	.1257	7.953	1.008	.9922	1.4457	50
20	.1280	.1276	7.834	.1287	7.770	1.008	.9918	1.4428	40
30	.1309	.1305	7.661	.1317	7.596	1.009	.9914	1.4399	30
40	.1338	.1334	7.496	.1346	7.429	1.009	.9911	1.4370	20
50	.1367	.1363	7.337	.1376	7.269	1.009	.9907	1.4341	10
8° 00′	.1396	.1392	7.185	.1405	7.115	1.010	.9903	1.4312	82° 00′
10	.1425	.1421	7.040	.1435	6.968	1.010	.9899	1.4283	50
20	.1454	.1449	6.900	.1465	6.827	1.011	.9894	1.4254	40
30	.1484	.1478	6.765	.1495	6.691	1.011	.9890	1.4224	30
40	.1513	.1507	6.636	.1524	6.561	1.012	.9886	1.4195	20
50	.1542	.1536	6.512	.1554	6.435	1.012	.9881	1.4166	10
9° 00′	.1571	.1564	6.392	.1584	6.314	1.012	.9877	1.4137	81° 00′
10	.1600	.1593	277	.1614	197	1.013	.9872	1.4108	50
20	.1629	.1622	166	.1644	084	1.013	.9868	1.4079	40
30	.1658	.1650	6.059	.1673	5.976	1.014	.9863	1.4050	30
40	.1687	.1679	5.955	.1703	871	1.014	.9858	1.4021	20
50	.1716	.1708	855	.1733	769	1.015	.9853	1.3992	10
10° 00′	.1745	.1736	5.759	1.763	5.671	1.015	.9848	1.3963	80° 00′
10	.1774	.1765	665	.1793	576	1.016	.9843	1.3934	50
20	.1804	.1794	575	.1823	485	1.016	.9838	1.3904	40
30	.1833	.1822	5.487	.1853	5.396	1.017	.9833	1.3875	30
40	.1862	.1851	403	.1883	309	1.018	.9827	1.3846	20
50	.1891	.1880	320	.1914	226	1.018	.9822	1.3817	10
11° 00′	.1920	.1908	5.241	.1944	5.145	1.019	.9816	1.3788	79° 00′
10	.1949	.1937	164	.1974	066	1.019	.9811	1.3759	50
20	.1978	.1965	089	.2004	4.989	1.020	.9805	1.3730	40
30	.2007	.1994	5.016	.2035	4.915	1.020	.9799	1.3701	30
40	.2036	.2022	4.945	.2065	843	1.021	.9793	1.3672	20
50	.2065	.2051	876	.2095	773	1.022	.9787	1.3643	10
12° 00′	.2094	.2079	4.810	.2126	4.705	1.022	.9781	1.3614	78° 00′
		cos θ	sec θ	cot θ	tan θ	csc θ	sin θ	θ radians	θ degrees

TABLE I, CONTINUED
TRIGONOMETRIC FUNCTIONS, VALUES IN DEGREES

θ degrees	θ radians	sin θ	csc θ	tan θ	cot θ	sec θ	cos θ		
12° 00′	.2094	.2079	4.810	.2126	4.705	1.022	.9781	1.3614	**78° 00′**
10	.2123	.2108	4.745	.2156	4.638	1.023	.9775	1.3584	50
20	.2153	.2136	4.682	.2186	4.574	1.024	.9769	1.3555	40
30	.2182	.2164	4.620	.2217	4.511	1.024	.9763	1.3526	30
40	.2211	.2193	4.560	.2247	4.449	1.025	.9757	1.3497	20
50	.2240	.2221	4.502	.2278	4.390	1.026	.9750	1.3468	10
13° 00′	.2269	.2250	4.445	.2309	4.331	1.026	.9744	1.3439	**77° 00′**
10	.2298	.2278	4.390	.2339	4.275	1.027	.9737	1.3410	50
20	.2327	.2306	4.336	.2370	4.219	1.028	.9730	1.3381	40
30	.2356	.2334	4.284	.2401	4.165	1.028	.9724	1.3352	30
40	.2385	.2363	4.232	.2432	4.113	1.029	.9717	1.3323	20
50	.2414	.2391	4.182	.2462	4.061	1.030	.9710	1.3294	10
14° 00′	.2443	.2419	4.134	.2493	4.011	1.031	.9703	1.3265	**76° 00′**
10	.2473	.2447	4.086	.2524	3.962	1.031	.9696	1.3235	50
20	.2502	.2476	4.039	.2555	3.914	1.032	.9689	1.3206	40
30	.2531	.2504	3.994	.2586	3.867	1.033	.9681	1.3177	30
40	.2560	.2532	3.950	.2617	3.821	1.034	.9674	1.3148	20
50	.2589	.2560	3.906	.2648	3.776	1.034	.9667	1.3119	10
15° 00′	.2618	.2588	3.864	.2679	3.732	1.035	.9659	1.3090	**75° 00′**
10	.2647	.2616	3.822	.2711	3.689	1.036	.9652	1.3061	50
20	.2676	.2644	3.782	.2742	3.647	1.037	.9644	1.3032	40
30	.2705	.2672	3.742	.2773	3.606	1.038	.9636	1.3003	30
40	.2734	.2700	3.703	.2805	3.566	1.039	.9628	1.2974	20
50	.2763	.2728	3.665	.2836	3.526	1.039	.9621	1.2945	10
16° 00′	.2793	.2756	3.628	.2867	3.487	1.040	.9613	1.2915	**74° 00′**
10	.2822	.2784	3.592	.2899	3.450	1.041	.9605	1.2886	50
20	.2851	.2812	3.556	.2931	3.412	1.042	.9596	1.2857	40
30	.2880	.2840	3.521	.2962	3.376	1.043	.9588	1.2828	30
40	.2909	.2868	3.487	.2994	3.340	1.044	.9580	1.2799	20
50	.2938	.2896	3.453	.3026	3.305	1.045	.9572	1.2770	10
17° 00′	.2967	.2924	3.420	.3057	3.271	1.046	.9563	1.2741	**73° 00′**
10	.2996	.2952	3.388	.3089	3.237	1.047	.9555	1.2712	50
20	.3025	.2979	3.357	.3121	3.204	1.048	.9546	1.2683	40
30	.3054	.3007	3.326	.3153	3.172	1.048	.9537	1.2654	30
40	.3083	.3035	3.295	.3185	3.140	1.049	.9528	1.2625	20
50	.3113	.3062	3.265	.3217	3.108	1.050	.9520	1.2595	10
18° 00′	.3142	.3090	3.236	.3249	3.078	1.051	.9511	1.2566	**72° 00′**
		cos θ	sec θ	cot θ	tan θ	csc θ	sin θ	θ radians	θ degrees

TABLE I, CONTINUED
TRIGONOMETRIC FUNCTIONS, VALUES IN DEGREES

θ degrees	θ radians	sin θ	csc θ	tan θ	cot θ	sec θ	cos θ		
18° 00′	.3142	.3090	3.236	.3249	3.078	1.051	.9511	1.2566	**72° 00′**
10	.3171	.3118	3.207	.3281	3.047	1.052	.9502	1.2537	50
20	.3200	.3145	3.179	.3314	3.018	1.053	.9492	1.2508	40
30	.3229	.3173	3.152	.3346	2.989	1.054	.9483	1.2479	30
40	.3258	.3201	3.124	.3378	2.960	1.056	.9474	1.2450	20
50	.3287	.3228	3.098	.3411	2.932	1.057	.9465	1.2421	10
19° 00′	.3316	.3256	3.072	.3443	2.904	1.058	.9455	1.2392	**71° 00′**
10	.3345	.3283	3.046	.3476	2.877	1.059	.9446	1.2363	50
20	.3374	.3311	3.021	.3508	2.850	1.060	.9436	1.2334	40
30	.3403	.3338	2.996	.3541	2.824	1.061	.9426	1.2305	30
40	.3432	.3365	2.971	.3574	2.798	1.062	.9417	1.2275	20
50	.3462	.3393	2.947	.3607	2.773	1.063	.9407	1.2246	10
20° 00′	.3491	.3420	2.924	.3640	2.747	1.064	.9397	1.2217	**70° 00′**
10	.3520	.3448	2.901	.3673	2.723	1.065	.9387	1.2188	50
20	.3549	.3475	2.878	.3706	2.699	1.066	.9377	1.2159	40
30	.3578	.3502	2.855	.3739	2.675	1.068	.9367	1.2130	30
40	.3607	.3529	2.833	.3772	2.651	1.069	.9356	1.2101	20
50	.3636	.3557	2.812	.3805	2.628	1.070	.9346	1.2072	10
21° 00′	.3665	.3584	2.790	.3839	2.605	1.071	.9336	1.2043	**69° 00′**
10	.3694	.3611	2.769	.3872	2.583	1.072	.9325	1.2014	50
20	.3723	.3638	2.749	.3906	2.560	1.074	.9315	1.2985	40
30	.3752	.3665	2.729	.3939	2.539	1.075	.9304	1.1956	30
40	.3782	.3692	2.709	.3973	2.517	1.076	.9293	1.1926	20
50	.3811	.3719	2.689	.4006	2.496	1.077	.9283	1.1897	10
22° 00′	.3840	.3746	2.669	.4040	2.475	1.079	.9272	1.1868	**68° 00′**
10	.3869	.3773	2.650	.4074	2.455	1.080	.9261	1.1839	50
20	.3898	.3800	2.632	.4108	2.434	1.081	.9250	1.1810	40
30	.3927	.3827	2.613	.4142	2.414	1.082	.9239	1.1781	30
40	.3956	.3854	2.595	.4176	2.394	1.084	.9228	1.1752	20
50	.3985	.3881	2.577	.4210	2.375	1.085	.9216	1.1723	10
23° 00′	.4014	.3907	2.559	.4245	2.356	1.086	.9205	1.1694	**67° 00′**
10	.4043	.3934	2.542	.4279	2.337	1.088	.9194	1.1665	50
20	.4072	.3961	2.525	.4314	2.318	1.089	.9182	1.1636	40
30	.4102	.3987	2.508	.4348	2.300	1.090	.9171	1.1606	30
40	.4131	.4014	2.491	.4383	2.282	1.092	.9159	1.1577	20
50	.4160	.4041	2.475	.4417	2.264	1.093	.9147	1.1548	10
24° 00′	.4189	.4067	2.459	.4452	2.246	1.095	.9135	1.1519	**66° 00′**
		cos θ	sec θ	cot θ	tan θ	csc θ	sin θ	θ radians	θ degrees

TABLE I, CONTINUED
TRIGONOMETRIC FUNCTIONS, VALUES IN DEGREES

θ degrees	θ radians	sin θ	csc θ	tan θ	cot θ	sec θ	cos θ		
24° 00′	.4189	.4067	2.459	.4452	2.246	1.095	.9135	1.1519	66° 00′
10	.4218	.4094	2.443	.4487	2.229	1.096	.9124	1.1490	50
20	.4247	.4120	2.427	.4522	2.211	1.097	.9112	1.1461	40
30	.4276	.4147	2.411	.4557	2.194	1.099	.9100	1.1432	30
40	.4305	.4173	2.396	.4592	2.177	1.100	.9088	1.1403	20
50	.4334	.4200	2.381	.4628	2.161	1.102	.9075	1.1374	10
25° 00′	.4363	.4226	2.366	.4663	2.145	1.103	.9063	1.1345	65° 00′
10	.4392	.4253	2.352	.4699	2.128	1.105	.9051	1.1316	50
20	.4422	.4279	2.337	.4734	2.112	1.106	.9038	1.1286	40
30	.4451	.4305	2.323	.4770	2.097	1.108	.9026	1.1257	30
40	.4480	.4331	2.309	.4806	2.081	1.109	.9013	1.1228	20
50	.4509	.4358	2.295	.4841	2.066	1.111	.9001	1.1199	10
26° 00′	.4538	.4384	2.281	.4877	2.050	1.113	.8988	1.1170	64° 00′
10	.4567	.4410	2.268	.4913	2.035	1.114	.8975	1.1141	50
20	.4596	.4436	2.254	.4950	2.020	1.116	.8962	1.1112	40
30	.4625	.4462	2.241	.4986	2.006	1.117	.8949	1.1083	30
40	.4654	.4488	2.228	.5022	1.991	1.119	.8936	1.1054	20
50	.4683	.4514	2.215	.5059	1.977	1.121	.8923	1.1025	10
27° 00′	.4712	.4540	2.203	.5095	1.963	1.122	.8910	1.0996	63° 00′
10	.4741	.4566	2.190	.5132	949	1.124	.8897	1.0966	50
20	.4771	.4592	2.178	.5169	935	1.126	.8884	1.0937	40
30	.4800	.4617	2.166	.5206	1.921	1.127	.8870	1.0908	30
40	.4829	.4643	2.154	.5243	907	1.129	.8857	1.0879	20
50	.4858	.4669	2.142	.5280	894	1.131	.8843	1.0850	10
28° 00′	.4887	.4695	2.130	.5317	1.881	1.133	.8829	1.0821	62° 00′
10	.4916	.4720	2.118	.5354	868	1.134	.8816	1.0792	50
20	.4945	.4746	2.107	.5392	855	1.136	.8802	1.0763	40
30	.4974	.4772	2.096	.5430	1.842	1.138	.8788	1.0734	30
40	.5003	.4797	2.085	.5467	829	1.140	.8774	1.0705	20
50	.5032	.4823	2.074	.5505	816	1.142	.8760	1.0676	10
29° 00′	.5061	.4848	2.063	.5543	1.804	1.143	.8746	1.0647	61° 00′
10	.5091	.4874	2.052	.5581	792	1.145	.8732	1.0617	50
20	.5120	.4899	2.041	.5619	780	1.147	.8718	1.0588	40
30	.5149	.4924	2.031	.5658	1.767	1.149	.8704	1.0559	30
40	.5178	.4950	2.020	.5696	756	1.151	.8689	1.0530	20
50	.5207	.4975	2.010	.5735	744	1.153	.8675	1.0501	10
30° 00′	.5236	.5000	2.000	.5774	1.732	1.155	.8660	1.0472	60° 00′
		cos θ	sec θ	cot θ	tan θ	csc θ	sin θ	θ radians	θ degrees

TABLE I, CONTINUED
TRIGONOMETRIC FUNCTIONS, VALUES IN DEGREES

θ degrees	θ radians	sin θ	csc θ	tan θ	cot θ	sec θ	cos θ		
30° 00′	.5236	.5000	2.000	.5774	1.732	1.155	.8660	1.0472	**60° 00′**
10	.5265	.5025	1.990	.5812	1.720	1.157	.8646	1.0443	50
20	.5294	.5050	1.980	.5851	1.709	1.159	.8631	1.0414	40
30	.5323	.5075	1.970	.5890	1.698	1.161	.8616	1.0385	30
40	.5352	.5100	1.961	.5930	1.686	1.163	.8601	1.0356	20
50	.5381	.5125	1.951	.5969	1.675	1.165	.8587	1.0327	10
31° 00′	.5411	.5150	1.942	.6009	1.664	1.167	.8572	1.0297	**59° 00′**
10	.5440	.5175	1.932	.6048	1.653	1.169	.8557	1.0268	50
20	.5469	.5200	1.923	.6088	1.643	1.171	.8542	1.0239	40
30	.5498	.5225	1.914	.6128	1.632	1.173	.8526	1.0210	30
40	.5527	.5250	1.905	.6168	1.621	1.175	.8511	1.0181	20
50	.5556	.5275	1.896	.6208	1.611	1.177	.8496	1.0152	10
32° 00′	.5585	.5299	1.887	.6249	1.600	1.179	.8480	1.0123	**58° 00′**
10	.5614	.5324	1.878	.6289	1.590	1.181	.8465	1.0094	50
20	.5643	.5348	1.870	.6330	1.580	1.184	.8450	1.0065	40
30	.5672	.5373	1.861	.6371	1.570	1.186	.8434	1.0036	30
40	.5701	.5398	1.853	.6412	1.560	1.188	.8418	1.0007	20
50	.5730	.5422	1.844	.6453	1.550	1.190	.8403	.9977	10
33° 00′	.5760	.5446	1.836	.6494	1.540	1.192	.8387	.9948	**57° 00′**
10	.5789	.5471	1.828	.6536	1.530	1.195	.8371	.9919	50
20	.5818	.5495	1.820	.6577	1.520	1.197	.8355	.9890	40
30	.5847	.5519	1.812	.6619	1.511	1.199	.8339	.9861	30
40	.5876	.5544	1.804	.6661	1.501	1.202	.8323	.9832	20
50	.5905	.5568	1.796	.6703	1.492	1.204	.8307	.9803	10
34° 00′	.5934	.5592	1.788	.6745	1.483	1.206	.8290	.9774	**56° 00′**
10	.5963	.5616	1.781	.6787	1.473	1.209	.8274	.9745	50
20	.5992	.5640	1.773	.6830	1.464	1.211	.8258	.9716	40
30	.6021	.5664	1.766	.6873	1.455	1.213	.8241	.9687	30
40	.6050	.5688	1.758	.6916	1.446	1.216	.8225	.9657	20
50	.6080	.5712	1.751	.6959	1.437	1.218	.8208	.9628	10
35° 00′	.6109	.5736	1.743	.7002	1.428	1.221	.8192	.9599	**55° 00′**
10	.6138	.5760	1.736	.7046	1.419	1.223	.8175	.9570	50
20	.6167	.5783	1.729	.7089	1.411	1.226	.8158	.9541	40
30	.6196	.5807	1.722	.7133	1.402	1.228	.8141	.9512	30
40	.6225	.5831	1.715	.7177	1.393	1.231	.8124	.9483	20
50	.6254	.5854	1.708	.7221	1.385	1.233	.8107	.9454	10
36° 00′	.6283	.5878	1.701	.7265	1.376	1.236	.8090	.9425	**54° 00′**
		cos θ	sec θ	cot θ	tan θ	csc θ	sin θ	θ radians	θ degrees

Table I, Continued
Trigonometric Functions, Values in Degrees

θ degrees	θ radians	sin θ	csc θ	tan θ	cot θ	sec θ	cos θ		
36° 00′	.6283	.5878	1.701	.7265	1.376	1.236	.8090	.9425	**54° 00′**
10	.6312	.5901	1.695	.7310	1.368	1.239	.8073	.9396	50
20	.6341	.5925	1.688	.7355	1.360	1.241	.8056	.9367	40
30	.6370	.5948	1.681	.7400	1.351	1.244	.8039	.9338	30
40	.6400	.5972	1.675	.7445	1.343	1.247	.8021	.9308	20
50	.6429	.5995	1.668	.7490	1.335	1.249	.8004	.9279	10
37° 00′	.6458	.6018	1.662	.7536	1.327	1.252	.7986	.9250	**53° 00′**
10	.6487	.6041	1.655	.7581	1.319	1.255	.7969	.9221	50
20	.6516	.6065	1.649	.7627	1.311	1.258	.7951	.9192	40
30	.6545	.6088	1.643	.7673	1.303	1.260	.7934	.9163	30
40	.6574	.6111	1.636	.7720	1.295	1.263	.7916	.9134	20
50	.6603	.6134	1.630	.7766	1.288	1.266	.7898	.9105	10
38° 00′	.6632	.6157	1.624	.7813	1.280	1.269	.7880	.9076	**52° 00′**
10	.6661	.6180	1.618	.7860	1.272	1.272	.7862	.9047	50
20	.6690	.6202	1.612	.7907	1.265	1.275	.7844	.9018	40
30	.6720	.6225	1.606	.7954	1.257	1.278	.7826	.8988	30
40	.6749	.6248	1.601	.8002	1.250	1.281	.7808	.8959	20
50	.6778	.6271	1.595	.8050	1.242	1.284	.7790	.8930	10
39° 00′	.6807	.6293	1.589	.8098	1.235	1.287	.7771	.8901	**51° 00′**
10	.6836	.6316	1.583	.8146	1.228	1.290	.7753	.8872	50
20	.6865	.6338	1.578	.8195	1.220	1.293	.7735	.8843	40
30	.6894	.6361	1.572	.8243	1.213	1.296	.7716	.8814	30
40	.6923	.6383	1.567	.8292	1.206	1.299	.7698	.8785	20
50	.6952	.6406	1.561	.8342	1.199	1.302	.7679	.8756	10
40° 00′	.6981	.6428	1.556	.8391	1.192	1.305	.7660	.8727	**50° 00′**
10	.7010	.6450	1.550	.8441	1.185	1.309	.7642	.8698	50
20	.7039	.6472	1.545	.8491	1.178	1.312	.7623	.8668	40
30	.7069	.6494	1.540	.8541	1.171	1.315	.7604	.8639	30
40	.7098	.6517	1.535	.8591	1.164	1.318	.7585	.8610	20
50	.7127	.6539	1.529	.8642	1.157	1.322	.7566	.8581	10
41° 00′	.7156	.6561	1.524	.8693	1.150	1.325	.7547	.8552	**49° 00′**
10	.7185	.6583	1.519	.8744	1.144	1.328	.7528	.8523	50
20	.7214	.6604	1.514	.8796	1.137	1.332	.7509	.8494	40
30	.7243	.6626	1.509	.8847	1.130	1.335	.7490	.8465	30
40	.7272	.6648	1.504	.8899	1.124	1.339	.7470	.8436	20
50	.7301	.6670	1.499	.8952	1.117	1.342	.7451	.8407	10
42° 00′	.7330	.6691	1.494	.9004	1.111	1.346	.7431	.8378	**48° 00′**
		cos θ	sec θ	cot θ	tan θ	csc θ	sin θ	θ radians	θ degrees

TABLE I, CONTINUED
TRIGONOMETRIC FUNCTIONS, VALUES IN DEGREES

θ degrees	θ radians	sin θ	csc θ	tan θ	cot θ	sec θ	cos θ		
42° 00′	.7330	.6691	1.494	.9004	1.111	1.346	.7431	.8378	**48° 00′**
10	.7359	.6713	1.490	.9057	1.104	1.349	.7412	.8348	50
20	.7389	.6734	1.485	.9110	1.098	1.353	.7392	.8319	40
30	.7418	.6756	1.480	.9163	1.091	1.356	.7373	.8290	30
40	.7447	.6777	1.476	.9217	1.085	1.360	.7353	.8261	20
50	.7476	.6799	1.471	.9271	1.079	1.364	.7333	.8232	10
43° 00′	.7505	.6820	1.466	.9325	1.072	1.367	.7314	.8203	**47° 00′**
10	.7534	.6841	1.462	.9380	1.066	1.371	.7294	.8174	50
20	.7563	.6862	1.457	.9435	1.060	1.375	.7274	.8145	40
30	.7592	.6884	1.453	.9490	1.054	1.379	.7254	.8116	30
40	.7621	.6905	1.448	.9545	1.048	1.382	.7234	.8087	20
50	.7650	.6926	1.444	.9601	1.042	1.386	.7214	.8058	10
44° 00′	.7679	.6947	1.440	.9657	1.036	1.390	.7193	.8029	**46° 00′**
10	.7709	.6967	1.435	.9713	1.030	1.394	.7173	.7999	50
20	.7738	.6988	1.431	.9770	1.024	1.398	.7153	.7970	40
30	.7767	.7009	1.427	.9827	1.018	1.402	.7133	.7941	30
40	.7796	.7030	1.423	.9884	1.012	1.406	.7112	.7912	20
50	.7825	.7050	1.418	.9942	1.006	1.410	.7092	.7883	10
45° 00′	.7854	.7071	1.414	1.000	1.000	1.414	.7071	.7854	**45° 00′**
		cos θ	sec θ	cot θ	tan θ	csc θ	sin θ	θ radians	θ degrees

Table II
Trigonometric Functions, Values in Radians

θ radians	θ degrees	sin θ	csc θ	tan θ	cot θ	sec θ	cos θ
0.00	0° 00′	.0000	undefined	.0000	undefined	1.000	1.0000
.01	0° 34′	.0100	100.0	.0100	100.0	1.000	1.0000
.02	1° 09′	.0200	50.00	.0200	49.99	1.000	.9998
.03	1° 43′	.0300	33.34	.0300	33.32	1.000	.9996
.04	2° 18′	.0400	25.01	.0400	24.99	1.001	.9992
0.05	2° 52′	.0500	20.01	.0500	19.98	1.001	.9988
.06	3° 26′	.0600	16.68	.0601	16.65	1.002	.9982
.07	4° 01′	.0699	14.30	.0701	14.26	1.002	.9976
.08	4° 35′	.0799	12.51	.0802	12.47	1.003	.9968
.09	5° 09′	.0899	11.13	.0902	11.08	1.004	.9960
0.10	5° 44′	.0998	10.02	.1003	9.967	1.005	.9950
.11	6° 18′	.1098	9.109	.1104	9.054	1.006	.9940
.12	6° 53′	.1197	8.353	.1206	8.293	1.007	.9928
.13	7° 27′	.1296	7.714	.1307	7.649	1.009	.9916
.14	8° 01′	.1395	7.166	.1409	7.096	1.010	.9902
0.15	8° 36′	.1494	6.692	.1511	6.617	1.011	.9888
.16	9° 10′	.1593	6.277	.1614	6.197	1.013	.9872
.17	9° 44′	.1692	5.911	.1717	5.826	1.015	.9856
.18	10° 19′	.1790	5.586	.1820	5.495	1.016	.9838
.19	10° 53′	.1889	5.295	.1923	5.200	1.018	.9820
0.20	11° 28′	.1987	5.033	.2027	4.933	1.020	.9801
.21	12° 02′	.2085	4.797	.2131	4.692	1.022	.9780
.22	12° 36′	.2182	4.582	.2236	4.472	1.025	.9759
.23	13° 11′	.2280	4.386	.2341	4.271	1.027	.9737
.24	13° 45′	.2377	4.207	.2447	4.086	1.030	.9713
0.25	14° 19′	.2474	4.042	.2553	3.916	1.032	.9689
.26	14° 54′	.2571	3.890	.2660	3.759	1.035	.9664
.27	15° 28′	.2667	3.749	.2768	3.613	1.038	.9638
.28	16° 03′	.2764	3.619	.2876	3.478	1.041	.9611
.29	16° 37′	.2860	3.497	.2984	3.351	1.044	.9582
0.30	17° 11′	.2955	3.384	.3093	3.233	1.047	.9553
.31	17° 46′	.3051	3.278	.3203	3.122	1.050	.9523
.32	18° 20′	.3146	3.179	.3314	3.018	1.053	.9492
.33	18° 54′	.3240	3.086	.3425	2.920	1.057	.9460
.34	19° 29′	.3335	2.999	.3537	2.827	1.061	.9428
0.35	20° 03′	.3429	2.916	.3650	2.740	1.065	.9394
.36	20° 38′	.3523	2.839	.3764	2.657	1.068	.9359
.37	21° 12′	.3616	2.765	.3879	2.578	1.073	.9323
.38	21° 46′	.3709	2.696	.3994	2.504	1.077	.9287
.39	22° 21′	.3802	2.630	.4111	2.433	1.081	.9249
0.40	22° 55′	.3894	2.568	.4228	2.365	1.086	.9211
θ radians	θ degrees	sin θ	csc θ	tan θ	cot θ	sec θ	cos θ

TABLE II, CONTINUED
TRIGONOMETRIC FUNCTIONS, VALUES IN RADIANS

θ radians	θ degrees	sin θ	csc θ	tan θ	cot θ	sec θ	cos θ
0.40	22° 55′	.3894	2.568	.4228	2.365	1.086	.9211
.41	23° 29′	.3986	2.509	.4346	2.301	1.090	.9171
.42	24° 04′	.4078	2.452	.4466	2.239	1.095	.9131
.43	24° 38′	.4169	2.399	.4586	2.180	1.100	.9090
.44	25° 13′	.4259	2.348	.4708	2.124	1.105	.9048
0.45	25° 47′	.4350	2.299	.4831	2.070	1.111	.9004
.46	26° 21′	.4439	2.253	.4954	2.018	1.116	.8961
.47	26° 56′	.4529	2.208	.5080	1.969	1.122	.8916
.48	27° 30′	.4618	2.166	.5206	1.921	1.127	.8870
.49	28° 04′	.4706	2.125	.5334	1.875	1.133	.8823
0.50	28° 39′	.4794	2.086	.5463	1.830	1.139	.8776
.51	29° 13′	.4882	2.048	.5594	1.788	1.146	.8727
.52	29° 48′	.4969	2.013	.5726	1.747	1.152	.8678
.53	30° 22′	.5055	1.978	.5859	1.707	1.159	.8628
.54	30° 56′	.5141	1.945	.5994	1.668	1.166	.8577
0.55	31° 31′	.5227	1.913	.6131	1.631	1.173	.8525
.56	32° 05′	.5312	1.883	.6269	1.595	1.180	.8473
.57	32° 40′	.5396	1.853	.6410	1.560	1.188	.8419
.58	33° 14′	.5480	1.825	.6552	1.526	1.196	.8365
.59	33° 48′	.5564	1.797	.6696	1.494	1.203	.8309
0.60	34° 23′	.5646	1.771	.6841	1.462	1.212	.8253
.61	34° 57′	.5729	1.746	.6989	1.431	1.220	.8196
.62	35° 31′	.5810	1.721	.7139	1.401	1.229	.8139
.63	36° 06′	.5891	1.697	.7291	1.372	1.238	.8080
.64	36° 40′	.5972	1.674	.7445	1.343	1.247	.8021
0.65	37° 15′	.6052	1.652	.7602	1.315	1.256	.7961
.66	37° 49′	.6131	1.631	.7761	1.288	1.266	.7900
.67	38° 23′	.6210	1.610	.7923	1.262	1.276	.7838
.68	38° 58′	.6288	1.590	.8087	1.237	1.286	.7776
.69	39° 32′	.6365	1.571	.8253	1.212	1.297	.7712
0.70	40° 06′	.6442	1.552	.8423	1.187	1.307	.7648
.71	40° 41′	.6518	1.534	.8595	1.163	1.319	.7584
.72	41° 15′	.6594	1.517	.8771	1.140	1.330	.7518
.73	41° 50′	.6669	1.500	.8949	1.117	1.342	.7452
.74	42° 24′	.6743	1.483	.9131	1.095	1.354	.7385
0.75	42° 58′	.6816	1.467	.9316	1.073	1.367	.7317
.76	43° 33′	.6889	1.452	.9505	1.052	1.380	.7248
.77	44° 07′	.6961	1.436	.9697	1.031	1.393	.7179
.78	44° 41′	.7033	1.422	.9893	1.011	1.407	.7109
.79	45° 16′	.7104	1.408	1.009	.9908	1.421	.7038
0.80	45° 50′	.7174	1.394	1.030	.9712	1.435	.6967
θ radians	θ degrees	sin θ	csc θ	tan θ	cot θ	sec θ	cos θ

Table II, Continued
Trigonometric Functions, Values in Radians

θ radians	θ degrees	sin θ	csc θ	tan θ	cot θ	sec θ	cos θ
0.80	45° 50′	.7174	1.394	1.030	.9712	1.435	.6967
.81	46° 25′	.7243	1.381	1.050	.9520	1.450	.6895
.82	46° 59′	.7311	1.368	1.072	.9331	1.466	.6822
.83	47° 33′	.7379	1.355	1.093	.9146	1.482	.6749
.84	48° 08′	.7446	1.343	1.116	.8964	1.498	.6675
0.85	48° 42′	.7513	1.331	1.138	.8785	1.515	.6600
.86	49° 16′	.7578	1.320	1.162	.8609	1.533	.6524
.87	49° 51′	.7643	1.308	1.185	.8437	1.551	.6448
.88	50° 25′	.7707	1.297	1.210	.8267	1.569	.6372
.89	51° 00′	.7771	1.287	1.235	.8100	1.589	.6294
0.90	51° 34′	.7833	1.277	1.260	.7936	1.609	.6216
.91	52° 08′	.7895	1.267	1.286	.7774	1.629	.6137
.92	52° 43′	.7956	1.257	1.313	.7615	1.651	.6058
.93	53° 17′	.8016	1.247	1.341	.7458	1.673	.5978
.94	53° 51′	.8076	1.238	1.369	.7303	1.696	.5898
0.95	54° 26′	.8134	1.229	1.398	.7151	1.719	.5817
.96	55° 00′	.8192	1.221	1.428	.7001	1.744	.5735
.97	55° 35′	.8249	1.212	1.459	.6853	1.769	.5653
.98	56° 09′	.8305	1.204	1.491	.6707	1.795	.5570
.99	56° 43′	.8360	1.196	1.524	.6563	1.823	.5487
1.00	57° 18′	.8415	1.188	1.557	.6421	1.851	.5403
1.01	57° 52′	.8468	1.181	1.592	.6281	1.880	.5319
1.02	58° 27′	.8521	1.174	1.628	.6142	1.911	.5234
1.03	59° 01′	.8573	1.166	1.665	.6005	1.942	.5148
1.04	59° 35′	.8624	1.160	1.704	.5870	1.975	.5062
1.05	60° 10′	.8674	1.153	1.743	.5736	2.010	.4976
1.06	60° 44′	.8724	1.146	1.784	.5604	2.046	.4889
1.07	61° 18′	.8772	1.140	1.827	.5473	2.083	.4801
1.08	61° 53′	.8820	1.134	1.871	.5344	2.122	.4713
1.09	62° 27′	.8866	1.128	1.917	.5216	2.162	.4625
1.10	63° 02′	.8912	1.122	1.965	.5090	2.205	.4536
1.11	63° 36′	.8957	1.116	2.014	.4964	2.249	.4447
1.12	64° 10′	.9001	1.111	2.066	.4840	2.295	.4357
1.13	64° 45′	.9044	1.106	2.120	.4718	2.344	.4267
1.14	65° 19′	.9086	1.101	2.176	.4596	2.395	.4176
1.15	65° 53′	.9128	1.096	2.234	.4475	2.448	.4085
1.16	66° 28′	.9168	1.091	2.296	.4356	2.504	.3993
1.17	67° 02′	.9208	1.086	2.360	.4237	2.563	.3902
1.18	67° 37′	.9246	1.082	2.247	.4120	2.625	.3809
1.19	68° 11′	.9284	1.077	2.498	.4003	2.691	.3717
1.20	68° 45′	.9320	1.073	2.572	.3888	2.760	.3624
θ radians	θ degrees	sin θ	csc θ	tan θ	cot θ	sec θ	cos θ

TABLE II, CONTINUED
TRIGONOMETRIC FUNCTIONS, VALUES IN RADIANS

θ radians	θ degrees	sin θ	csc θ	tan θ	cot θ	sec θ	cos θ
1.20	68° 45′	.9320	1.073	2.572	.3888	2.760	.3624
1.21	69° 20′	.9356	1.069	2.650	.3773	2.833	.3530
1.22	69° 54′	.9391	1.065	2.733	.3659	2.910	.3436
1.23	70° 28′	.9425	1.061	2.820	.3546	2.992	.3342
1.24	71° 03′	.9458	1.057	2.912	.3434	3.079	.3248
1.25	71° 37′	.9490	1.054	3.010	.3323	3.171	.3153
1.26	72° 12′	.9521	1.050	3.113	.3212	3.270	.3058
1.27	72° 46′	.9551	1.047	3.224	.3102	3.375	.2963
1.28	73° 20′	.9580	1.044	3.341	.2993	3.488	.2867
1.29	73° 55′	.9608	1.041	3.467	.2884	3.609	.2771
1.30	74° 29′	.9636	1.038	3.602	.2776	3.738	.2675
1.31	75° 03′	.9662	1.035	3.747	.2669	3.878	.2579
1.32	75° 38′	.9687	1.032	3.903	.2562	4.029	.2482
1.33	76° 12′	.9711	1.030	4.072	.2456	4.193	.2385
1.34	76° 47′	.9735	1.027	4.256	.2350	4.372	.2288
1.35	77° 21′	.9757	1.025	4.455	.2245	4.566	.2190
1.36	77° 55′	.9779	1.023	4.673	.2140	4.779	.2092
1.37	78° 30′	.9799	1.021	4.913	.2035	5.014	.1994
1.38	79° 04′	.9819	1.018	5.177	.1931	5.273	.1896
1.39	79° 38′	.9837	1.017	5.471	.1828	5.561	.1798
1.40	80° 13′	.9854	1.015	5.798	.1725	5.883	.1700
1.41	80° 47′	.9871	1.013	6.165	.1622	6.246	.1601
1.42	81° 22′	.9887	1.011	6.581	.1519	6.657	.1502
1.43	81° 56′	.9901	1.010	7.055	.1417	7.126	.1403
1.44	82° 30′	.9915	1.009	7.602	.1315	7.667	.1304
1.45	83° 05′	.9927	1.007	8.238	.1214	8.299	.1205
1.46	83° 39′	.9939	1.006	8.989	.1113	9.044	.1106
1.47	84° 13′	.9949	1.005	9.887	.1011	9.938	.1006
1.48	84° 48′	.9959	1.004	10.98	.0910	11.03	.0907
1.49	85° 22′	.9967	1.003	12.35	.0810	12.39	.0807
1.50	85° 57′	.9975	1.003	14.10	.0709	14.14	.0707
1.51	86° 31′	.9982	1.002	16.43	.0609	16.46	.0608
1.52	87° 05′	.9987	1.001	19.67	.0508	19.69	.0508
1.53	87° 40′	.9992	1.001	24.50	.0408	24.52	.0408
1.54	88° 14′	.9995	1.000	32.46	.0308	32.48	.0308
1.55	88° 49′	.9998	1.000	48.08	.0208	48.09	.0208
1.56	89° 23′	.9999	1.000	92.62	.0108	92.63	.0108
1.57	89° 57′	1.0000	1.000	1256	.0008	1256	.0008
θ radians	θ degrees	sin θ	csc θ	tan θ	cot θ	sec θ	cos θ

Table III
Logarithms, Base 10

n	0	1	2	3	4	5	6	7	8	9
1.0	+0.0000	.0043	.0086	.0128	.0170	.0212	.0253	.0294	.0334	.0374
1.1	.0414	.0453	.0492	.0531	.0569	.0607	.0645	.0682	.0719	.0755
1.2	.0792	.0828	.0864	.0899	.0934	.0969	.1004	.1038	.1072	.1106
1.3	.1139	.1173	.1206	.1239	.1271	.1303	.1335	.1367	.1399	.1430
1.4	.1461	.1492	.1523	.1553	.1584	.1614	.1644	.1673	.1703	.1732
1.5	.1761	.1790	.1818	.1847	.1875	.1903	.1931	.1959	.1987	.2014
1.6	.2041	.2068	.2095	.2122	.2148	.2175	.2201	.2227	.2253	.2279
1.7	.2304	.2330	.2355	.2380	.2405	.2430	.2455	.2480	.2504	.2529
1.8	.2553	.2577	.2601	.2625	.2648	.2672	.2695	.2718	.2742	.2765
1.9	.2788	.2810	.2833	.2856	.2878	.2900	.2923	.2945	.2967	.2989
2.0	.3010	.3032	.3054	.3075	.3096	.3118	.3139	.3160	.3181	.3201
2.1	.3222	.3243	.3263	.3284	.3304	.3324	.3345	.3365	.3385	.3404
2.2	.3424	.3444	.3464	.3483	.3502	.3522	.3541	.3560	.3579	.3598
2.3	.3617	.3636	.3655	.3674	.3692	.3711	.3729	.3747	.3766	.3784
2.4	.3802	.3820	.3838	.3856	.3874	.3892	.3909	.3927	.3945	.3962
2.5	.3979	.3997	.4014	.4031	.4048	.4065	.4082	.4099	.4116	.4133
2.6	.4150	.4166	.4183	.4200	.4216	.4232	.4249	.4265	.4281	.4298
2.7	.4314	.4330	.4346	.4362	.4378	.4393	.4409	.4425	.4440	.4456
2.8	.4472	.4487	.4502	.4518	.4533	.4548	.4564	.4579	.4594	.4609
2.9	.4624	.4639	.4654	.4669	.4683	.4698	.4713	.4728	.4742	.4757
3.0	.4771	.4786	.4800	.4814	.4829	.4843	.4857	.4871	.4886	.4900
3.1	.4914	.4928	.4942	.4955	.4969	.4983	.4997	.5011	.5024	.5038
3.2	.5051	.5065	.5079	.5092	.5105	.5119	.5132	.5145	.5159	.5172
3.3	.5185	.5198	.5211	.5224	.5237	.5250	.5263	.5276	.5289	.5302
3.4	.5315	.5328	.5340	.5353	.5366	.5378	.5391	.5403	.5416	.5428
3.5	.5441	.5453	.5465	.5478	.5490	.5502	.5514	.5527	.5539	.5551
3.6	.5563	.5575	.5587	.5599	.5611	.5623	.5635	.5647	.5658	.5670
3.7	.5682	.5694	.5705	.5717	.5729	.5740	.5752	.5763	.5775	.5786
3.8	.5798	.5809	.5821	.5832	.5843	.5855	.5866	.5877	.5888	.5899
3.9	.5911	.5922	.5933	.5944	.5955	.5966	.5977	.5988	.5999	.6010
4.0	.6021	.6031	.6042	.6053	.6064	.6075	.6085	.6096	.6107	.6117
4.1	.6128	.6138	.6149	.6160	.6170	.6180	.6191	.6201	.6212	.6222
4.2	.6232	.6243	.6253	.6263	.6274	.6284	.6294	.6304	.6314	.6325
4.3	.6335	.6345	.6355	.6365	.6375	.6385	.6395	.6405	.6415	.6425
4.4	.6435	.6444	.6454	.6464	.6474	.6484	.6493	.6503	.6513	.6522
4.5	.6532	.6542	.6551	.6561	.6571	.6580	.6590	.6599	.6609	.6618
4.6	.6628	.6637	.6646	.6656	.6665	.6675	.6684	.6693	.6702	.6712
4.7	.6721	.6730	.6739	.6749	.6758	.6767	.6776	.6785	.6794	.6803
4.8	.6812	.6821	.6830	.6839	.6848	.6857	.6866	.6875	.6884	.6893
4.9	.6902	.6911	.6920	.6928	.6937	.6946	.6955	.6964	.6972	.6981
5.0	.6990	.6998	.7007	.7016	.7024	.7033	.7042	.7050	.7059	.7067
5.1	.7076	.7084	.7093	.7101	.7110	.7118	.7126	.7135	.7143	.7152
5.2	.7160	.7168	.7177	.7185	.7193	.7202	.7210	.7218	.7226	.7235
5.3	.7243	.7251	.7259	.7267	.7275	.7284	.7292	.7300	.7308	.7316
5.4	.7324	.7332	.7340	.7348	.7356	.7364	.7372	.7380	.7388	.7396
n	0	1	2	3	4	5	6	7	8	9

Table III, Continued
Logarithms, Base 10

n	0	1	2	3	4	5	6	7	8	9
5.5	.7404	.7412	.7419	.7427	.7435	.7443	.7451	.7459	.7466	.7474
5.6	.7482	.7490	.7497	.7505	.7513	.7520	.7528	.7536	.7543	.7551
5.7	.7559	.7566	.7574	.7582	.7589	.7597	.7604	.7612	.7619	.7627
5.8	.7634	.7642	.7649	.7657	.7664	.7672	.7679	.7686	.7694	.7701
5.9	.7709	.7716	.7723	.7731	.7738	.7745	.7752	.7760	.7767	.7774
6.0	.7782	.7789	.7796	.7803	.7810	.7818	.7825	.7832	.7839	.7846
6.1	.7853	.7860	.7868	.7875	.7882	.7889	.7896	.7903	.7910	.7917
6.2	.7924	.7931	.7938	.7945	.7952	.7959	.7966	.7973	.7980	.7987
6.3	.7993	.8000	.8007	.8014	.8021	.8028	.8035	.8041	.8048	.8055
6.4	.8062	.8069	.8075	.8082	.8089	.8096	.8102	.8109	.8116	.8122
6.5	.8129	.8136	.8142	.8149	.8156	.8162	.8169	.8176	.8182	.8189
6.6	.8195	.8202	.8209	.8215	.8222	.8228	.8235	.8241	.8248	.8254
6.7	.8261	.8267	.8274	.8280	.8287	.8293	.8299	.8306	.8312	.8319
6.8	.8325	.8331	.8338	.8344	.8351	.8357	.8363	.8370	.8376	.8382
6.9	.8388	.8395	.8401	.8407	.8414	.8420	.8426	.8432	.8439	.8445
7.0	.8451	.8457	.8463	.8470	.8476	.8482	.8488	.8494	.8500	.8506
7.1	.8513	.8519	.8525	.8531	.8537	.8543	.8549	.8555	.8561	.8567
7.2	.8573	.8579	.8585	.8591	.8597	.8603	.8609	.8615	.8621	.8627
7.3	.8633	.8639	.8645	.8651	.8657	.8663	.8669	.8675	.8681	.8686
7.4	.8692	.8698	.8704	.8710	.8716	.8722	.8727	.8733	.8739	.8745
7.5	.8751	.8756	.8762	.8768	.8774	.8779	.8785	.8791	.8797	.8802
7.6	.8808	.8814	.8820	.8825	.8831	.8837	.8842	.8848	.8854	.8859
7.7	.8865	.8871	.8876	.8882	.8887	.8893	.8899	.8904	.8910	.8915
7.8	.8921	.8927	.8932	.8938	.8943	.8949	.8954	.8960	.8965	.8971
7.9	.8976	.8982	.8987	.8993	.8998	.9004	.9009	.9015	.9020	.9025
8.0	.9031	.9036	.9042	.9047	.9053	.9058	.9063	.9069	.9074	.9079
8.1	.9085	.9090	.9096	.9101	.9106	.9112	.9117	.9122	.9128	.9133
8.2	.9138	.9143	.9149	.9154	.9159	.9165	.9170	.9175	.9180	.9186
8.3	.9191	.9196	.9201	.9206	.9212	.9217	.9222	.9227	.9232	.9238
8.4	.9243	.9248	.9253	.9258	.9263	.9269	.9274	.9279	.9284	.9289
8.5	.9294	.9299	.9304	.9309	.9315	.9320	.9325	.9330	.9335	.9340
8.6	.9345	.9350	.9355	.9360	.9365	.9370	.9375	.9380	.9385	.9390
8.7	.9395	.9400	.9405	.9410	.9415	.9420	.9425	.9430	.9435	.9440
8.8	.9445	.9450	.9455	.9460	.9465	.9469	.9474	.9479	.9484	.9489
8.9	.9494	.9499	.9504	.9509	.9513	.9518	.9523	.9528	.9533	.9538
9.0	.9542	.9547	.9552	.9557	.9562	.9566	.9571	.9576	.9581	.9586
9.1	.9590	.9595	.9600	.9605	.9609	.9614	.9619	.9624	.9628	.9633
9.2	.9638	.9643	.9647	.9652	.9657	.9661	.9666	.9671	.9675	.9680
9.3	.9685	.9689	.9694	.9699	.9703	.9708	.9713	.9717	.9722	.9727
9.4	.9731	.9736	.9741	.9745	.9750	.9754	.9759	.9763	.9768	.9773
9.5	.9777	.9782	.9786	.9791	.9795	.9800	.9805	.9809	.9814	.9818
9.6	.9823	.9827	.9832	.9836	.9841	.9845	.9850	.9854	.9859	.9863
9.7	.9868	.9872	.9877	.9881	.9886	.9890	.9894	.9899	.9903	.9908
9.8	.9912	.9917	.9921	.9926	.9930	.9934	.9939	.9943	.9948	.9952
9.9	.9956	.9961	.9965	.9969	.9974	.9978	.9983	.9987	.9991	.9996
n	0	1	2	3	4	5	6	7	8	9

TABLE IV
LOGARITHMS OF TRIGONOMETRIC FUNCTIONS

Attach − 10 to logarithms obtained from this table.

θ degrees	log sin θ	log csc θ	log tan θ	log cot θ	log sec θ	log cos θ	
0° 00′	undefined	undefined	undefined	undefined	10.0000	10.0000	90° 00′
10′	7.4637	12.5363	7.4637	12.5363	10.0000	10.0000	50′
20′	7.7648	12.2352	7.7648	12.2352	10.0000	10.0000	40′
30′	7.9408	12.0592	7.9409	12.0591	10.0000	10.0000	30′
40′	8.0658	11.9342	8.0658	11.9342	10.0000	10.0000	20′
50′	8.1627	11.8373	8.1627	11.8373	10.0000	10.0000	10′
1° 00′	8.2419	11.7581	8.2419	11.7581	10.0001	9.9999	89° 00′
10′	8.3088	11.6912	8.3089	11.6911	10.0001	9.9999	50′
20′	8.3668	11.6332	8.3669	11.6331	10.0001	9.9999	40′
30′	8.4179	11.5821	8.4181	11.5819	10.0001	9.9999	30′
40′	8.4637	11.5363	8.4638	11.5362	10.0002	9.9998	20′
50′	8.5050	11.4950	8.5053	11.4947	10.0002	9.9998	10′
2° 00′	8.5428	11.4572	8.5431	11.4569	10.0003	9.9997	88° 00′
10′	8.5776	11.4224	8.5779	11.4221	10.0003	9.9997	50′
20′	8.6097	11.3903	8.6101	11.3899	10.0004	9.9996	40′
30′	8.6397	11.3603	8.6401	11.3599	10.0004	9.9996	30′
40′	8.6677	11.3323	8.6682	11.3318	10.0005	9.9995	20′
50′	8.6940	11.3060	8.6945	11.3055	10.0005	9.9995	10′
3° 00′	8.7188	11.2812	8.7194	11.2806	10.0006	9.9994	87° 00′
10′	8.7423	11.2577	8.7429	11.2571	10.0007	9.9993	50′
20′	8.7645	11.2355	8.7652	11.2348	10.0007	9.9993	40′
30′	8.7857	11.2143	8.7865	11.2135	10.0008	9.9992	30′
40′	8.8059	11.1941	8.8067	11.1933	10.0009	9.9991	20′
50′	8.8251	11.1749	8.8261	11.1739	10.0010	9.9990	10′
4° 00′	8.8436	11.1564	8.8446	11.1554	10.0011	9.9989	86° 00′
10′	8.8613	11.1387	8.8624	11.1376	10.0011	9.9989	50′
20′	8.8783	11.1217	8.8795	11.1205	10.0012	9.9988	40′
30′	8.8946	11.1054	8.8960	11.1040	10.0013	9.9987	30′
40′	8.9104	11.0896	8.9118	11.0882	10.0014	9.9986	20′
50′	8.9256	11.0744	8.9272	11.0728	10.0015	9.9985	10′
5° 00′	8.9403	11.0597	8.9420	11.0580	10.0017	9.9983	85° 00′
10′	8.9545	11.0455	8.9563	11.0437	10.0018	9.9982	50′
20′	8.9682	11.0318	8.9701	11.0299	10.0019	9.9981	40′
30′	8.9816	11.0184	8.9836	11.0164	10.0020	9.9980	30′
40′	8.9945	11.0055	8.9966	11.0034	10.0021	9.9979	20′
50′	9.0070	10.9930	9.0093	10.9907	10.0023	9.9977	10′
6° 00′	9.0192	10.9808	9.0216	10.9784	10.0024	9.9976	84° 00′
	log cos θ	log sec θ	log cot θ	log tan θ	log csc θ	log sin θ	θ degrees

Table IV, Continued
Logarithms of Trigonometric Functions

Attach − 10 to logarithms obtained from this table.

θ degrees	log sin θ	log csc θ	log tan θ	log cot θ	log sec θ	log cos θ	
6° 00′	9.0192	10.9808	9.0216	10.9784	10.0024	9.9976	**84° 00′**
10′	9.0311	10.9689	9.0336	10.9664	10.0025	9.9975	50′
20′	9.0426	10.9574	9.0453	10.9547	10.0027	9.9973	40′
30′	9.0539	10.9461	9.0567	10.9433	10.0028	9.9972	30′
40′	9.0648	10.9352	9.0678	10.9322	10.0029	9.9971	20′
50′	9.0755	10.9245	9.0786	10.9214	10.0031	9.9969	10′
7° 00′	9.0859	10.9141	9.0891	10.9109	10.0032	9.9968	**83° 00′**
10′	9.0961	10.9039	9.0995	10.9005	10.0034	9.9966	50′
20′	9.1060	10.8940	9.1096	10.8904	10.0036	9.9964	40′
30′	9.1157	10.8843	9.1194	10.8806	10.0037	9.9963	30′
40′	9.1252	10.8748	9.1291	10.8709	10.0039	9.9961	20′
50′	9.1345	10.8655	9.1385	10.8615	10.0041	9.9959	10′
8° 00′	9.1436	10.8564	9.1478	10.8522	10.0042	9.9958	**82° 00′**
10′	9.1525	10.8475	9.1569	10.8431	10.0044	9.9956	50′
20′	9.1612	10.8388	9.1658	10.8342	10.0046	9.9954	40′
30′	9.1697	10.8303	9.1745	10.8255	10.0048	9.9952	30′
40′	9.1781	10.8219	9.1831	10.8169	10.0050	9.9950	20′
50′	9.1863	10.8137	9.1915	10.8085	10.0052	9.9948	10′
9° 00′	9.1943	10.8057	9.1997	10.8003	10.0054	9.9946	**81° 00′**
10′	9.2022	10.7978	9.2078	10.7922	10.0056	9.9944	50′
20′	9.2100	10.7900	9.2158	10.7842	10.0058	9.9942	40′
30′	9.2176	10.7824	9.2236	10.7764	10.0060	9.9940	30′
40′	9.2251	10.7749	9.2313	10.7687	10.0062	9.9938	20′
50′	9.2324	10.7676	9.2389	10.7611	10.0064	9.9936	10′
10° 00′	9.2397	10.7603	9.2463	10.7537	10.0066	9.9934	**80° 00′**
10′	9.2468	10.7532	9.2536	10.7464	10.0069	9.9931	50′
20′	9.2538	10.7462	9.2609	10.7391	10.0071	9.9929	40′
30′	9.2606	10.7394	9.2680	10.7320	10.0073	9.9927	30′
40′	9.2674	10.7326	9.2750	10.7250	10.0076	9.9924	20′
50′	9.2740	10.7260	9.2819	10.7181	10.0078	9.9922	10′
11° 00′	9.2806	10.7194	9.2887	10.7113	10.0081	9.9919	**79° 00′**
10′	9.2870	10.7130	9.2953	10.7047	10.0083	9.9917	50′
20′	9.2934	10.7066	9.3020	10.6980	10.0086	9.9914	40′
30′	9.2997	10.7003	9.3085	10.6915	10.0088	9.9912	30′
40′	9.3058	10.6942	9.3149	10.6851	10.0091	9.9909	20′
50′	9.3119	10.6881	9.3212	10.6788	10.0093	9.9907	10′
12° 00′	9.3179	10.6821	9.3275	10.6725	10.0096	9.9904	**78° 00′**
	log cos θ	log sec θ	log cot θ	log tan θ	log csc θ	log sin θ	θ degrees

Table IV, Continued
Logarithms of Trigonometric Functions

Attach − 10 to logarithms obtained from this table.

θ degrees	log sin θ	log csc θ	log tan θ	log cot θ	log sec θ	log cos θ	
12° 00′	9.3179	10.6821	9.3275	10.6725	10.0096	9.9904	**78° 00′**
10′	9.3238	10.6762	9.3336	10.6664	10.0099	9.9901	50′
20′	9.3296	10.6704	9.3397	10.6603	10.0101	9.9899	40′
30′	9.3353	10.6647	9.3458	10.6542	10.0104	9.9896	30′
40′	9.3410	10.6590	9.3517	10.6483	10.0107	9.9893	20′
50′	9.3466	10.6534	9.3576	10.6424	10.0110	9.9890	10′
13° 00′	9.3521	10.6479	9.3634	10.6366	10.0113	9.9887	**77° 00′**
10′	9.3575	10.6425	9.3691	10.6309	10.0116	9.9884	50′
20′	9.3629	10.6371	9.3748	10.6252	10.0119	9.9881	40′
30′	9.3682	10.6318	9.3804	10.6196	10.0122	9.9878	30′
40′	9.3734	10.6266	9.3859	10.6141	10.0125	9.9875	20′
50′	9.3786	10.6214	9.3914	10.6086	10.0128	9.9872	10′
14° 00′	9.3837	10.6163	9.3968	10.6032	10.0131	9.9869	**76° 00′**
10′	9.3887	10.6113	9.4021	10.5979	10.0134	9.9866	50′
20′	9.3937	10.6063	9.4074	10.5926	10.0137	9.9863	40′
30′	9.3986	10.6014	9.4127	10.5873	10.0141	9.9859	30′
40′	9.4035	10.5965	9.4178	10.5822	10.0144	9.9856	20′
50′	9.4083	10.5917	9.4230	10.5770	10.0147	9.9853	10′
15° 00′	9.4130	10.5870	9.4281	10.5719	10.0151	9.9849	**75° 00′**
10′	9.4177	10.5823	9.4331	10.5669	10.0154	9.9846	50′
20′	9.4223	10.5777	9.4381	10.5619	10.0157	9.9843	40′
30′	9.4269	10.5731	9.4330	10.5570	10.0161	9.9839	30′
40′	9.4314	10.5686	9.4479	10.5521	10.0164	9.9836	20′
50′	9.4359	10.5641	9.4527	10.5473	10.0168	9.9832	10′
16° 00′	9.4403	10.5597	9.4575	10.5425	10.0172	9.9828	**74° 00′**
10′	9.4447	10.5553	9.4622	10.5378	10.0175	9.9825	50′
20′	9.4491	10.5509	9.4669	10.5331	10.0179	9.9821	40′
30′	9.4533	10.5467	9.4716	10.5284	10.0183	9.9817	30′
40′	9.4576	10.5424	9.4762	10.5238	10.0186	9.9814	20′
50′	9.4618	10.5382	9.4808	10.5192	10.0190	9.9810	10′
17° 00′	9.4659	10.5341	9.4853	10.5147	10.0194	9.9806	**73° 00′**
10′	9.4700	10.5300	9.4898	10.5102	10.0198	9.9802	50′
20′	9.4741	10.5259	9.4943	10.5057	10.0202	9.9798	40′
30′	9.4781	10.5219	9.4987	10.5013	10.0206	9.9794	30′
40′	9.4821	10.5179	9.5031	10.4969	10.0210	9.9790	20′
50′	9.4861	10.5139	9.5075	10.4925	10.0214	9.9786	10′
18° 00′	9.4900	10.5100	9.5118	10.4882	10.0218	9.9782	**72° 00′**
	log cos θ	log sec θ	log cot θ	log tan θ	log csc θ	log sin θ	θ degrees

Table IV, Continued
Logarithms of Trigonometric Functions

Attach -10 to logarithms obtained from this table.

θ degrees	log sin θ	log csc θ	log tan θ	log cot θ	log sec θ	log cos θ	
18° 00'	9.4900	10.5100	9.5118	10.4882	10.0218	9.9782	**72° 00'**
10'	9.4939	10.5061	9.5161	10.4839	10.0222	9.9778	50'
20'	9.4977	10.5023	9.5203	10.4797	10.0226	9.9774	40'
30'	9.5015	10.4985	9.5245	10.4755	10.0230	9.9770	30'
40'	9.5052	10.4948	9.5287	10.4713	10.0235	9.9765	20'
50'	9.5090	10.4910	9.5329	10.4671	10.0239	9.9761	10'
19° 00'	9.5126	10.4874	9.5370	10.4630	10.0243	9.9757	**71° 00'**
10'	9.5163	10.4837	9.5411	10.4589	10.0248	9.9752	50'
20'	9.5199	10.4801	9.5451	10.4549	10.0252	9.9748	40'
30'	9.5235	10.4765	9.5491	10.4509	10.0257	9.9743	30'
40'	9.5270	10.4730	9.5531	10.4469	10.0261	9.9739	20'
50'	9.5306	10.4694	9.5571	10.4429	10.0266	9.9734	10'
20° 00'	9.5341	10.4659	9.5611	10.4389	10.0270	9.9730	**70° 00'**
10'	9.5375	10.4625	9.5650	10.4350	10.0275	9.9725	50'
20'	9.5409	10.4591	9.5689	10.4311	10.0279	9.9721	40'
30'	9.5443	10.4557	9.5727	10.4273	10.0284	9.9716	30'
40'	9.5477	10.4523	9.5766	10.4234	10.0289	9.9711	20'
50'	9.5510	10.4490	9.5804	10.4196	10.0294	9.9706	10'
21° 00'	9.5543	10.4457	9.5842	10.4158	10.0298	9.9702	**69° 00'**
10'	9.5576	10.4424	9.5879	10.4121	10.0303	9.9797	50'
20'	9.5609	10.4391	9.5917	10.4083	10.0308	9.9692	40'
30'	9.5641	10.4359	9.5954	10.4046	10.0313	9.9687	30'
40'	9.5673	10.4327	9.5991	10.4009	10.0318	9.9682	20'
50'	9.5704	10.4296	9.6028	10.3972	10.0323	9.9677	10'
22° 00'	9.5736	10.4264	9.6064	10.3936	10.0328	9.9672	**68° 00'**
10'	9.5767	10.4233	9.6100	10.3900	10.0333	9.9667	50'
20'	9.5798	10.4202	9.6136	10.3864	10.0339	9.9661	40'
30'	9.5828	10.4172	9.6172	10.3828	10.0344	9.9656	30'
40'	9.5859	10.4141	9.6208	10.3792	10.0349	9.9651	20'
50'	9.5889	10.4111	9.6243	10.3757	10.0354	9.9646	10'
23° 00'	9.5919	10.4081	9.6279	10.3721	10.0360	9.9640	**67° 00'**
10'	9.5948	10.4052	9.6314	10.3686	10.0365	9.9635	50'
20'	9.5978	10.4022	9.6348	10.3652	10.0371	9.9629	40'
30'	9.6007	10.3993	9.6383	10.3617	10.0376	9.9624	30'
40'	9.6036	10.3964	9.6417	10.3583	10.0382	9.9618	20'
50'	9.6065	10.3935	9.6452	10.3548	10.0387	9.9613	10'
24° 00'	9.6093	10.3907	9.6486	10.3514	10.0393	9.9607	**66° 00'**
	log cos θ	log sec θ	log cot θ	log tan θ	log csc θ	log sin θ	θ degrees

TABLE IV, CONTINUED
LOGARITHMS OF TRIGONOMETRIC FUNCTIONS

Attach −10 to logarithms obtained from this table.

θ degrees	log sin θ	log csc θ	log tan θ	log cot θ	log sec θ	log cos θ	
24° 00′	9.6093	10.3907	9.6486	10.3514	10.0393	9.9607	**66° 00′**
10′	9.6121	10.3879	9.6520	10.3480	10.0398	9.9602	50′
20′	9.6149	10.3851	9.6553	10.3447	10.0404	9.9596	40′
30′	9.6177	10.3823	9.6587	10.3413	10.0410	9.9590	30′
40′	9.6205	10.3795	9.6620	10.3380	10.0416	9.9584	20′
50′	9.6232	10.3768	9.6654	10.3346	10.0421	9.9579	10′
25° 00′	9.6259	10.3741	9.6687	10.3313	10.0427	9.9573	**65° 00′**
10′	9.6286	10.3714	9.6720	10.3280	10.0433	9.9567	50′
20′	9.6313	10.3687	9.6752	10.3248	10.0439	9.9561	40′
30′	9.6340	10.3660	9.6785	10.3215	10.0445	9.9555	30′
40′	9.6366	10.3634	9.6817	10.3183	10.0451	9.9549	20′
50′	9.6392	10.3608	9.6850	10.3150	10.0457	9.9543	10′
26° 00′	9.6418	10.3582	9.6882	10.3118	10.0463	9.9537	**64° 00′**
10′	9.6444	10.3556	9.6914	10.3086	10.0470	9.9530	50′
20′	9.6470	10.3530	9.6946	10.3054	10.0476	9.9524	40′
30′	9.6495	10.3505	9.6977	10.3023	10.0482	9.9518	30′
40′	9.6521	10.3479	9.7009	10.2991	10.0488	9.9512	20′
50′	9.6546	10.3454	9.7040	10.2960	10.0495	9.9505	10′
27° 00′	9.6570	10.3430	9.7072	10.2928	10.0501	9.9499	**63° 00′**
10′	9.6595	10.3405	9.7103	10.2897	10.0508	9.9492	50′
20′	9.6620	10.3380	9.7134	10.2866	10.0514	9.9486	40′
30′	9.6644	10.3356	9.7165	10.2835	10.0521	9.9479	30′
40′	9.6668	10.3332	9.7196	10.2804	10.0527	9.9473	20′
50′	9.6692	10.3308	9.7226	10.2774	10.0534	9.9466	10′
28° 00′	9.6716	10.3284	9.7257	10.2743	10.0541	9.9459	**62° 00′**
10′	9.6740	10.3260	9.7287	10.2713	10.0547	9.9453	50′
20′	9.6763	10.3237	9.7317	10.2683	10.0554	9.9446	40′
30′	9.6787	10.3213	9.7348	10.2652	10.0561	9.9439	30′
40′	9.6810	10.3190	9.7378	10.2622	10.0568	9.9432	20′
50′	9.6833	10.3167	9.7408	10.2592	10.0575	9.9425	10′
29° 00′	9.6856	10.3144	9.7438	10.2562	10.0582	9.9418	**61° 00′**
10′	9.6878	10.3122	9.7467	10.2533	10.0589	9.9411	50′
20′	9.6901	10.3099	9.7497	10.2503	10.0596	9.9404	40′
30′	9.6923	10.3077	9.7526	10.2474	10.0603	9.9397	30′
40′	9.6946	10.3054	9.7556	10.2444	10.0610	9.9390	20′
50′	9.6968	10.3032	9.7585	10.2415	10.0617	9.9383	10′
30° 00′	9.6990	10.3010	9.7614	10.2386	10.0625	9.9375	**60° 00′**
	log cos θ	log sec θ	log cot θ	log tan θ	log csc θ	log sin θ	θ degrees

Table IV, Continued
Logarithms of Trigonometric Functions

Attach −10 to logarithms obtained from this table.

θ degrees	log sin θ	log csc θ	log tan θ	log cot θ	log sec θ	log cos θ	
30° 00′	9.6990	10.3010	9.7614	10.2386	10.0625	9.9375	**60° 00′**
10′	9.7012	10.2988	9.7644	10.2356	10.0632	9.9368	50′
20′	9.7033	10.2967	9.7673	10.2327	10.0639	9.9361	40′
30′	9.7055	10.2945	9.7701	10.2299	10.0647	9.9353	30′
40′	9.7076	10.2924	9.7730	10.2270	10.0654	9.9346	20′
50′	9.7097	10.2903	9.7759	10.2241	10.0662	9.9338	10′
31° 00′	9.7118	10.2882	9.7788	10.2212	10.0669	9.9331	**59° 00′**
10′	9.7139	10.2861	9.7816	10.2184	10.0677	9.9323	50′
20′	9.7160	10.2840	9.7845	10.2155	10.0685	9.9315	40′
30′	9.7181	10.2819	9.7873	10.2127	10.0692	9.9308	30′
40′	9.7201	10.2799	9.7902	10.2098	10.0700	9.9300	20′
50′	9.7222	10.2778	9.7930	10.2070	10.0708	9.9292	10′
32° 00′	9.7242	10.2758	9.7958	10.2042	10.0716	9.9284	**58° 00′**
10′	9.7262	10.2738	9.7986	10.2014	10.0724	9.9276	50′
20′	9.7282	10.2718	9.8014	10.1986	10.0732	9.9268	40′
30′	9.7302	10.2698	9.8042	10.1958	10.0740	9.9260	30′
40′	9.7322	10.2678	9.8070	10.1930	10.0748	9.9252	20′
50′	9.7342	10.2658	9.8097	10.1903	10.0756	9.9244	10′
33° 00′	9.7361	10.2639	9.8125	10.1875	10.0764	9.9236	**57° 00′**
10′	9.7380	10.2620	9.8153	10.1847	10.0772	9.9228	50′
20′	9.7400	10.2600	9.8180	10.1820	10.0781	9.9219	40′
30′	9.7419	10.2581	9.8208	10.1792	10.0789	9.9211	30′
40′	9.7438	10.2562	9.8235	10.1765	10.0797	9.9203	20′
50′	9.7457	10.2543	9.8263	10.1737	10.0806	9.9194	10′
34° 00′	9.7476	10.2524	9.8290	10.1710	10.0814	9.9186	**56° 00′**
10′	9.7494	10.2506	9.8317	10.1683	10.0823	9.9177	50′
20′	9.7513	10.2487	9.8344	10.1656	10.0831	9.9169	40′
30′	9.7531	10.2469	9.8371	10.1629	10.0840	9.9160	30′
40′	9.7550	10.2450	9.8398	10.1602	10.0849	9.9151	20′
50′	9.7568	10.2432	9.8425	10.1575	10.0858	9.9142	10′
35° 00′	9.7586	10.2414	9.8452	10.1548	10.0866	9.9134	**55° 00′**
10′	9.7604	10.2396	9.8479	10.1521	10.0875	9.9125	50′
20′	9.7622	10.2378	9.8506	10.1494	10.0884	9.9116	40′
30′	9.7640	10.2360	9.8533	10.1467	10.0893	9.9107	30′
40′	9.7657	10.2343	9.8559	10.1441	10.0902	9.9098	20′
50′	9.7675	10.2325	9.8586	10.1414	10.0911	9.9089	10′
36° 00′	9.7692	10.2308	9.8613	10.1387	10.0920	9.9080	**54° 00′**
	log cos θ	log sec θ	log cot θ	log tan θ	log csc θ	log sin θ	θ degrees

Table IV, Continued
Logarithms of Trigonometric Functions

Attach -10 to logarithms obtained from this table.

θ degrees	log sin θ	log csc θ	log tan θ	log cot θ	log sec θ	log cos θ	
36° 00′	9.7692	10.2308	9.8613	10.1387	10.0920	9.9080	**54° 00′**
10′	9.7710	10.2290	9.8639	10.1361	10.0930	9.9070	50′
20′	9.7727	10.2273	9.8666	10.1334	10.0939	9.9061	40′
30′	9.7744	10.2256	9.8692	10.1308	10.0948	9.9052	30′
40′	9.7761	10.2239	9.8718	10.1282	10.0958	9.9042	20′
50′	9.7778	10.2222	9.8745	10.1255	10.0967	9.9033	10′
37° 00′	9.7795	10.2205	9.8771	10.1229	10.0977	9.9023	**53° 00′**
10′	9.7811	10.2189	9.8797	10.1203	10.0986	9.9014	50′
20′	9.7828	10.2172	9.8824	10.1176	10.0996	9.9004	40′
30′	9.7844	10.2156	9.8850	10.1150	10.1005	9.8995	30′
40′	9.7861	10.2139	9.8876	10.1124	10.1015	9.8985	20′
50′	9.7877	10.2123	9.8902	10.1098	10.1025	9.8975	10′
38° 00′	9.7893	10.2107	9.8928	10.1072	10.1035	9.8965	**52° 00′**
10′	9.7910	10.2090	9.8954	10.1046	10.1045	9.8955	50′
20′	9.7926	10.2074	9.8980	10.1020	10.1055	9.8945	40′
30′	9.7941	10.2059	9.9006	10.0994	10.1065	9.8935	30′
40′	9.7957	10.2043	9.9032	10.0968	10.1075	9.8925	20′
50′	9.7973	10.2027	9.9058	10.0942	10.1085	9.8915	10′
39° 00′	9.7989	10.2011	9.9084	10.0916	10.1095	9.8905	**51° 00′**
10′	9.8004	10.1996	9.9110	10.0890	10.1105	9.8895	50′
20′	9.8020	10.1980	9.9135	10.0865	10.1116	9.8884	40′
30′	9.8035	10.1965	9.9161	10.0839	10.1126	9.8874	30′
40′	9.8050	10.1950	9.9187	10.0813	10.1136	9.8864	20′
50′	9.8066	10.1934	9.9212	10.0788	10.1147	9.8853	10′
40° 00′	9.8081	10.1919	9.9238	10.0762	10.1157	9.8843	**50° 00′**
10′	9.8096	10.1904	9.9264	10.0736	10.1168	9.8832	50′
20′	9.8111	10.1889	9.9289	10.0711	10.1179	9.8821	40′
30′	9.8125	10.1875	9.9315	10.0685	10.1190	9.8810	30′
40′	9.8140	10.1860	9.9341	10.0659	10.1200	9.8800	20′
50′	9.8155	10.1845	9.9366	10.0634	10.1211	9.8789	10′
41° 00′	9.8169	10.1831	9.9392	10.0608	10.1222	9.8778	**49° 00′**
10′	9.8184	10.1816	9.9417	10.0583	10.1233	9.8767	50′
20′	9.8198	10.1802	9.9443	10.0557	10.1244	9.8756	40′
30′	9.8213	10.1787	9.9468	10.0532	10.1255	9.8745	30′
40′	9.8227	10.1773	9.9494	10.0506	10.1267	9.8733	20′
50′	9.8241	10.1759	9.9519	10.0481	10.1278	9.8722	10′
42° 00′	9.8255	10.1745	9.9544	10.0456	10.1289	9.8711	**48° 00′**
	log cos θ	log sec θ	log cot θ	log tan θ	log csc θ	log sin θ	θ degrees

Table IV, Continued
Logarithms of Trigonometric Functions

Attach − 10 to logarithms obtained from this table.

θ degrees	log sin θ	log csc θ	log tan θ	log cot θ	log sec θ	log cos θ	
42° 00′	9.8255	10.1745	9.9544	10.0456	10.1289	9.8711	**48° 00′**
10′	9.8269	10.1731	9.9570	10.0430	10.1301	9.8699	50′
20′	9.8283	10.1717	9.9595	10.0405	10.1312	9.8688	40′
30′	9.8297	10.1703	9.9621	10.0379	10.1324	9.8676	30′
40′	9.8311	10.1689	9.9646	10.0354	10.1335	9.8665	20′
50′	9.8324	10.1676	9.9671	10.0329	10.1347	9.8653	10′
43° 00′	9.8338	10.1662	9.9697	10.0303	10.1359	9.8641	**47° 00′**
10′	9.8351	10.1649	9.9722	10.0278	10.1371	9.8629	50′
20′	9.8365	10.1635	9.9747	10.0253	10.1382	9.8618	40′
30′	9.8378	10.1622	9.9772	10.0228	10.1394	9.8606	30′
40′	9.8391	10.1609	9.9798	10.0202	10.1406	9.8594	20′
50′	9.8405	10.1595	9.9823	10.0177	10.1418	9.8582	10′
44° 00′	9.8418	10.1582	9.9848	10.0152	10.1431	9.8569	**46° 00′**
10′	9.8431	10.1569	9.9874	10.0126	10.1443	9.8557	50′
20′	9.8444	10.1556	9.9899	10.0101	10.1455	9.8545	40′
30′	9.8457	10.1543	9.9924	10.0076	10.1468	9.8532	30′
40′	9.8469	10.1531	9.9949	10.0051	10.1480	9.8520	20′
50′	9.8482	10.1518	9.9975	10.0025	10.1493	9.8507	10′
45° 00′	9.8495	10.1505	10.0000	10.0000	10.1505	9.8495	**45° 00′**
	log cos θ	log sec θ	log cot θ	log tan θ	log csc θ	log sin θ	θ degrees

Table V
Squares and Square Roots

n	n²	√n	n	n²	√n	n	n²	√n	n	n²	√n
0	0	0.000	50	2,500	7.071	100	10,000	10.000	150	22,500	12.247
1	1	1.000	51	2,601	7.141	101	10,201	10.050	151	22,801	12.288
2	4	1.414	52	2,704	7.211	102	10,404	10.100	152	23,104	12.329
3	9	1.732	53	2,809	7.280	103	10,609	10.149	153	23,409	12.369
4	16	2.000	54	2,916	7.348	104	10,816	10.198	154	23,716	12.410
5	25	2.236	55	3,025	7.416	105	11,025	10.247	155	24,025	12.450
6	36	2.449	56	3,136	7.483	106	11,236	10.296	156	24,336	12.490
7	49	2.646	57	3,249	7.550	107	11,449	10.344	157	24,649	12.530
8	64	2.828	58	3,364	7.616	108	11,664	10.392	158	24,964	12.570
9	81	3.000	59	3,481	7.681	109	11,881	10.440	159	25,281	12.610
10	100	3.162	60	3,600	7.746	110	12,100	10.488	160	25,600	12.649
11	121	3.317	61	3,721	7.810	111	12,321	10.536	161	25,921	12.689
12	144	3.464	62	3,844	7.874	112	12,544	10.583	162	26,244	12.728
13	169	3.606	63	3,969	7.937	113	12,769	10.630	163	26,569	12.767
14	196	3.742	64	4,096	8.000	114	12,996	10.677	164	26,896	12.806
15	225	3.873	65	4,225	8.062	115	13,225	10.724	165	27,225	12.845
16	256	4.000	66	4,356	8.124	116	13,456	10.770	166	27,556	12.884
17	289	4.123	67	4,489	8.185	117	13,689	10.817	167	27,889	12.923
18	324	4.243	68	4,624	8.246	118	13,924	10.863	168	28,224	12.961
19	361	4.359	69	4,761	8.307	119	14,161	10.909	169	28,561	13.000
20	400	4.472	70	4,900	8.367	120	14,400	10.954	170	28,900	13.038
21	441	4.583	71	5,041	8.426	121	14,641	11.000	171	29,241	13.077
22	484	4.690	72	5,184	8.485	122	14,884	11.045	172	29,584	13.115
23	529	4.796	73	5,329	8.544	123	15,129	11.091	173	29,929	13.153
24	576	4.899	74	5,476	8.602	124	15,376	11.136	174	30,276	13.191
25	625	5.000	75	5,625	8.660	125	15,625	11.180	175	30,625	13.229
26	676	5.099	76	5,776	8.718	126	15,876	11.225	176	30,976	13.266
27	729	5.196	77	5,929	8.775	127	16,129	11.269	177	31,329	13.304
28	784	5.292	78	6,084	8.832	128	16,384	11.314	178	31,684	13.342
29	841	5.385	79	6,241	8.888	129	16,641	11.358	179	32,041	13.379
30	900	5.477	80	6,400	8.944	130	16,900	11.402	180	32,400	13.416
31	961	5.568	81	6,561	9.000	131	17,161	11.446	181	32,761	13.454
32	1,024	5.657	82	6,724	9.055	132	17,424	11.489	182	33,124	13.491
33	1,089	5.745	83	6,889	9.110	133	17,689	11.533	183	33,489	13.528
34	1,156	5.831	84	7,056	9.165	134	17,956	11.576	184	33,856	13.565
35	1,225	5.916	85	7,225	9.220	135	18,225	11.619	185	34,225	13.601
36	1,296	6.000	86	7,396	9.274	136	18,496	11.662	186	34,596	13.638
37	1,369	6.083	87	7,569	9.327	137	18,769	11.705	187	34,969	13.675
38	1,444	6.164	88	7,744	9.381	138	19,044	11.747	188	35,344	13.711
39	1,521	6.245	89	7,921	9.434	139	19,321	11.790	189	35,721	13.748
40	1,600	6.325	90	8,100	9.487	140	19,600	11.832	190	36,100	13.784
41	1,681	6.403	91	8,281	9.539	141	19,881	11.874	191	36,481	13.820
42	1,764	6.481	92	8,464	9.592	142	20,164	11.916	192	36,864	13.856
43	1,849	6.557	93	8,649	9.644	143	20,449	11.958	193	37,249	13.892
44	1,936	6.633	94	8,836	9.659	144	20,736	12.000	194	37,636	13.928
45	2,025	6.708	95	9,025	9.747	145	21,025	12.042	195	38,025	13.964
46	2,116	6.782	96	9,216	9.798	146	21,316	12.083	196	38,416	14.000
47	2,209	6.856	97	9,409	9.849	147	21,609	12.124	197	38,809	14.036
48	2,304	6.928	98	9,604	9.899	148	21,904	12.166	198	39,204	14.071
49	2,401	7.000	99	9,801	9.950	149	22,201	12.207	199	39,601	14.107

Answers to Selected Problems

CHAPTER ONE

1.2
1. a. 0.5299   b. 0.5299   c. 0.5619
   d. 0.5619   e. 0.6450   f. 1.419   g. 0.1765
   h. 8.556
3. a. 31°00′   b. 59°00′   c. 46°10′
   d. 43°50′   e. 0°30′   f. 46°50′   g. 27°50′
   h. 69°40′
5. a. $2\sqrt{2}$ b. $\sin 45° = \cos 45° = \frac{1}{2}\sqrt{2}$, $\tan 45° = \cot 45° = 1$ c. $\sin 45° = \cos 45° \approx 0.7071$, $\tan 45° = \cot 45° = 1.000$ The values of $\tan 45°$ and $\cot 45°$ are exact.
7. a. 61°00′   b. 9.70   c. 17.5
9. a. 54°10′   b. 35°50′   c. 30.8
11. a. 6.40 m   b. 53°10′, 53°10′, 73°40′
    c. 30.7 sq m
13. 17.8 mm, 21.8 sq m   15. 31.9 mi

1.3
9. 1220 ft   11. 159 sq units
13. 238879 mi

1.4
1. a. 25°   b. 130°   c. 30°   d. 50°
3. a. 120°   b. 120°   c. 60°   d. 120°
5. a. 17   b. 25
7. 8   9. 5, 12, 13
11. 90 in., $(675/2)\sqrt{3}$ sq in.   13. 15, 5
15. $60\sqrt{2}$ in., 450 sq in.

1.5
1. a. {1, 2, 3, 4, 5}   b. {−3, −2, −1, 0, 1, 2, 3}
   c. {5, 22/7, 0, −2.1}
3. −155°, −95°, −35°, 25°, 85°, 145°, 205°
5. a. $6x^2 - 13x - 5$ b. $(\tan \alpha)^2 - 4$
   c. $16x^2 + 24x + 9$
7. a. $(4x + 5)(4x - 5)$ b. $(4x - 5)^2$
   c. $(3x - 5)(2x - 1)$
9. a. {−1, 4}   b. {1/3, 3}
11. a. {2.76, 7.24}   b. {0.20, 1.14}
13. a. 10, 3/4   b. $3\sqrt{2}$, 1
15. a. 45°   b. 149°

1.6
1. a. 430°, −290°, 790°   b. 230°, −490°, 590°
   c. 680°, −40°, 1040°
   d. $(17\pi/6)^R$, $(-7\pi/6)^R$, $(-19\pi/6)^R$
3. $(5\pi/12)^R$ 5. $(2\pi/3)^R$ 7. $(7\pi/6)^R$
9. $(\pi/2)^R$ 11. $(7\pi/3)^R$ 13. $(-5\pi/9)^R$
15. $(31\pi/36)^R$ 17. 210°   19. 165°
21. 360°   23. 135°   25. $(360/\pi)°$
27. −270°   29. 80°   31. 160°
33. $(\pi/4)^R$ 35. $(19\pi/12)^R$
37. $0° < \alpha < 45°$ 39. $0° < \alpha < 180°$
41. $-45° < \alpha < 0°$ 43. $90° < \alpha < 135°$
45. $\pi^R < \alpha < 2\pi^R$ 47. QII   49. QIII
51. QII

REVIEW EXERCISES
1. 24.0   2. 90.4   3. 21°50′   4. 60 ft
5. 5.88 in.   6. 0.554   7. 32.1   8. 4500 ft
10. 175 sq in.   11. a. $\sqrt{338} \approx 18.4$ b. (13, 0)
    c. $4\sqrt{13} \approx 14.4$ d. 90°   e. 67°20′
12. 19°50′   13. 79°20′   14. 46.3 sq cm
15. 36°20′   16. a. 1.17   b. 22°50′

CHAPTER TWO

2.1

Answers are presented in the following order: sin α, cos α, tan α, cot α, sec α, csc α.

1. $4/5, -3/5, -4/3, -3/4, -5/3, 5/4$
3. $-\frac{1}{2}\sqrt{2}, \frac{1}{2}\sqrt{2}, -1, -1, \sqrt{2}, -\sqrt{2}$
5. $-5/13, -12/13, 5/12, 12/5, -13/12, -13/5$
7. $3/5, 4/5, \mathbf{3/4}, 4/3, 5/4, 5/3$
9. $\sqrt{\mathbf{11/6}}, 5/6, \sqrt{11}/5, 5/\sqrt{11}, 6/5, 6/\sqrt{11}$
11. $8/17, 15/17, 8/15, 15/8, \mathbf{17/15}, 17/8$
13. $\mathbf{-4/5}, 3/5, -4/3, -3/4, 5/3, -5/4$
15. $-\sqrt{23}/12, -11/12, \sqrt{\mathbf{23/11}}, 11/\sqrt{23}, -12/11, -12/\sqrt{23}$
17. $1/2, -\sqrt{3}/2, -1/\sqrt{3}, -\sqrt{3}, -2/\sqrt{3}, 2$
19. $0, -1, 0,$ undefined, $-1,$ undefined
21. $\dfrac{\sin \alpha}{\cos \alpha} = \tan \alpha$

2.2

1. $1, 0, -, 0, -, 1$
3. $-1/2, -\sqrt{3}/2, 1/\sqrt{3}, \sqrt{3}, -2/\sqrt{3}, -2$
5. $\sqrt{3}/2, -1/2, -\sqrt{3}, -1/\sqrt{3}, -2, 2/\sqrt{3}$
7. $-\sqrt{3}/2, 1/2, -\sqrt{3}, -1/\sqrt{3}, 2, -2/\sqrt{3}$
9. $-\sqrt{2}/2, -\sqrt{2}/2, 1, 1, -\sqrt{2}, -\sqrt{2}$
11.

	0°	90°	180°	270°	360°
sin	0	1	0	−1	0
cos	1	0	−1	0	1
tan	0	—	0	—	0
cot	—	0	—	0	—
sec	1	—	−1	—	1
csc	—	1	—	−1	—

13. 1   15. 1   17. 1   19. 1
21. 1/2   23. −1/2   25. 1/4

2.3

1. 40°   3. 80°   5. 60°   7. 30°
9. 80°, 0.1763   11. 70°, −0.3420
13. 50°, 0.7660   15. 35°, −1.428
17. $1.54^R, -0.0308$ 19. $1.14^R, -2.176$
21. 30°50′, 149°10′   23. 121°, 301°
25. 25°50′, 334°10′   27. 0.73, 5.55
29. 2.22, 5.36
31. $46° + k \cdot 360°, 134° + k \cdot 360°$
33. $32° + k \cdot 180°$
35. $120° + k \cdot 360°, 240° + k \cdot 360°$
37. $1.17 + k \cdot \pi$ 39. $3.59 + k \cdot 2\pi, 5.83 + k \cdot 2\pi$

2.4

α	−1.20	−0.80	−0.40
α̲	1.20	0.80	0.40
cos α	0.36	0.70	0.92
cot α	−0.39	−0.97	−2.37

α	0.00	0.40	0.80	1.20
α̲	0.00	0.40	0.80	1.20
cos α	1.00	0.92	0.70	0.36
cot α	—	2.37	0.97	0.39

α	1.60	2.00	2.40	2.80
α̲	1.54	1.14	0.74	0.34
cos α	−0.03	−0.42	−0.74	−0.94
cot α	−0.03	−0.46	−1.10	−2.83

α	3.20	3.60	4.00	4.40
α̲	0.06	0.46	0.86	1.26
cos α	−0.99	−0.90	−0.65	−0.31
cot α	16.65	2.02	0.86	0.32

α	4.80	5.20	5.60	6.00
α̲	1.48	1.08	0.68	0.28
cos α	0.09	0.47	0.78	0.96
cot α	−0.09	−0.53	−1.24	−3.48

α	6.40	6.80	7.20	7.60
α̲	0.12	0.52	0.92	1.32
cos α	0.99	0.87	0.61	0.25
cot α	8.29	1.75	0.76	0.26

1. See Figure 2.4.1 on the next page.
3. See Figure 2.4.3 on the next page.
5. *a.* even   *b.* odd   *c.* neither   *d.* even
See Figure 2.4.5 on the next page.

ANSWERS TO SELECTED PROBLEMS

221

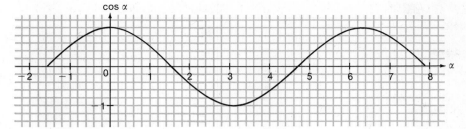

FIGURE 2.4.1

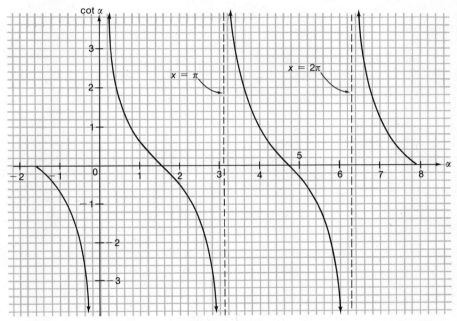

FIGURE 2.4.3

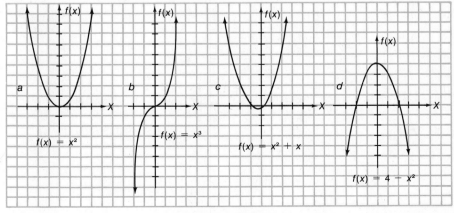

FIGURE 2.4.5

REVIEW EXERCISES

1. a. 0.5925 b. −0.8391 c. 0.3420
   d. −5.485 e. −2.411 f. −0.5854
   g. −0.7002 h. 0.1883
2. a. 0.8415 b. −0.5994 c. 0.2867
   d. 0.3659
3. a. 54°30′, 234°30′ b. 202°20′, 337°40′
   c. 120°40′, 239°20′ d. 52°30′, 232°30′
4. a. $0.66^R, 5.62^R$ b. $2.02^R, 5.16^R$
5. $x = 5, y = -12, r = 13$, so $\sin \alpha = -\dfrac{12}{13}$,
   $\cos \alpha = \dfrac{5}{13}, \tan \alpha = -\dfrac{12}{5}, \cot \alpha = -\dfrac{5}{12}$,
   $\sec \alpha = \dfrac{13}{5}, \csc \alpha = -\dfrac{13}{12}$.
6. Let $x = -2\sqrt{2}, r = 3$. Then $(-2\sqrt{2})^2 + y^2 = 3^2, y^2 = 9 - 8 = 1$, so $y = 1$.
   $\sin \alpha = \dfrac{1}{3}, \tan \alpha = -\dfrac{1}{2\sqrt{2}} = \dfrac{-\sqrt{2}}{4}$,
   $\cot \alpha = -2\sqrt{2}, \sec \alpha = \dfrac{3}{-2\sqrt{2}} = \dfrac{-3\sqrt{2}}{4}, \csc \alpha = 3$.
7. a. $\tan 300° = -\tan 60° = -\sqrt{3}$
   b. $\sin (5\pi/4)^R = -\sin (\pi/4)^R = -(1/2)\sqrt{2}$
   c. $\cos 390° = \cos 30° = (1/2)\sqrt{3}$
   d. $\cot (-\pi)$ is undefined.
8. a. Maximum values at (0, 1), (2π, 1); minimum value at (π, −1); zeros at π/2, 3π/2.

a
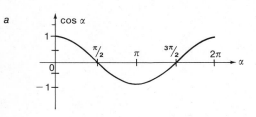

b. zeros at $0, \pi, 2\pi$; no maximum or minimum values; asymptotes are $x = \pi/2, x = 3\pi/2$.

b
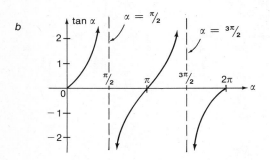

CHAPTER THREE

3.2
5. $(\sqrt{6} + \sqrt{2})/4$ 7. $(\sqrt{6} + \sqrt{2})/4$
9. 150°20′ 19. a. 4/5 b. −4/3
c. −33/56 d. −5/13 e. −12/13
f. 56/65 g. −63/65 h. 24/7
19. a. 4/5 b. −4/3 c. −33/56
d. −5/13 e. −12/13 f. 56/65
g. −63/65 h. 24/7
31. a. $2 \sin 3\alpha \cos \alpha$ b. $2 \cos (\pi/2) \cos (\pi/3)$
c. $2 \cos 5x \sin x$ d. $2 \sin (5\pi/6) \sin (\pi/2)$

3.3
1. $2 \sin 4\theta \cos 4\theta$ 3. $\tan 4\theta$ 5. $\cos 4\theta$
7. $|\cos 4\theta|$ 9. $(1/2)(1 - \cos 4\theta)$
11. $(1/2)\sqrt{2 - \sqrt{3}}$ 13. $2 - \sqrt{3}$
15. $2 - \sqrt{3}$ 17. $-\sqrt{3}/2$ 19. 24/25
21. −24/7 23. $-\sqrt{1/5}$ 25. −44/125
47. a. $\sin \theta = \sqrt{1/5}, \cos \theta = \sqrt{4/5}, \tan \theta = 1/2$

3.4
1. 120°, 240° 3. 60°, 120°, 240°, 300°
5. 30°, 150°, 270° 7. 75°, 165°, 255°, 345°
9. π/2, 7π/6, 3π/2, 11π/6 11. 0, 3π/2
13. $45° + k \cdot 180°$ 15. $90° + k \cdot 180°$
17. 0°, 30°, 150°, 180° 19. 0°
21. 45°, 108°30′, 225°, 288°30′
23. 216°50′, 323°10′
25. 45°, 123°40′, 225°, 303°40′
27. 60°, 180°, 300°
29. 43°40′, 136°20′, 196°50′, 343°10′
31. 61°10′, 140°20′, 241°50′, 320°20′
33. 57°15′, 122°45′, 237°15′, 302°45′
35. 0 37. −π/12 39. −π/2, 0, π/2

REVIEW EXERCISES
11. a. −12/13 b. −5/12 c. 8/17
    d. −15/17 e. 140/221 f. 120/119
    g. $2/\sqrt{13}$ h. −828/2035
12. a. $\frac{1}{4}(\sqrt{6} - \sqrt{2})$ b. $\frac{1}{2}\sqrt{2 - \sqrt{3}}$
13. a. {0°, 60°, 180°, 300°}
    b. {45°, 105°, 165°, 225°, 285°, 345°}
    c. {0°, 180°} d. {90°, 210°, 330°}
    e. {111°30′, 248°30′}
    f. {63°30′, 135°, 243°30′, 315°}

CHAPTER FOUR
4.1
1. *a.* 1   *b.* 0   *c.* −1   *d.* 0
3.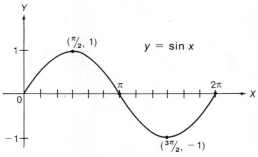
5.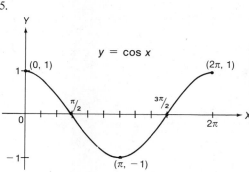
7.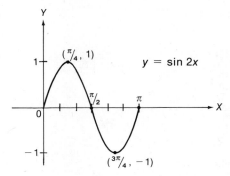
9.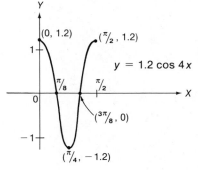
11.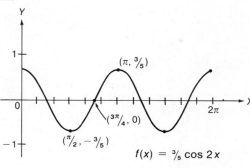
13.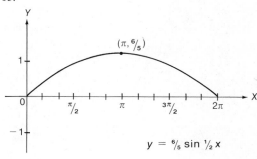
15.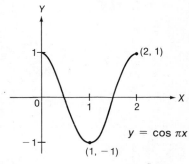
17.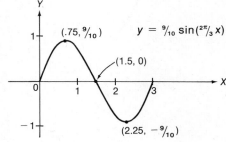

19.
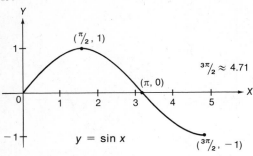

21.
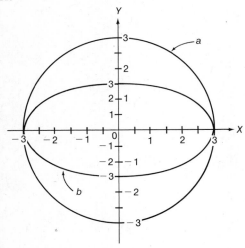

4.2
1. *a.* 0   *b.* 2   *c.* 0   *d.* 0
3.
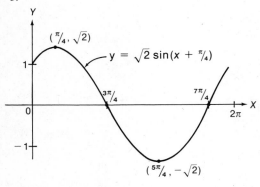

5.
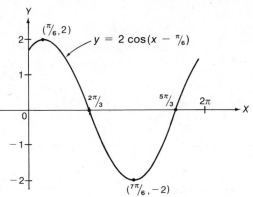

7.
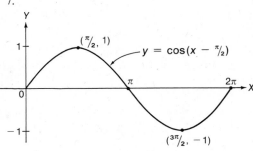

Note that $\cos(x - \pi/2) = \cos(\pi/2 - x) = \sin x$.

9.
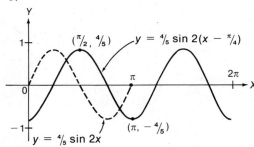

11.
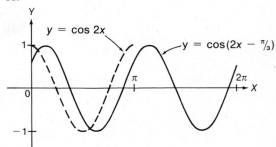

13.

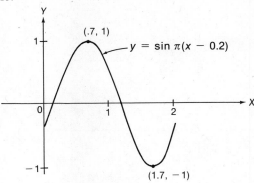

15.

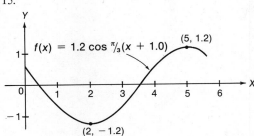

17. The graphs should be identical.

4.3
1. *a*. 2   *b*. $2\sqrt{2}$ *c*. undefined
   d. $-2\sqrt{2}$ *e*. -2
3.

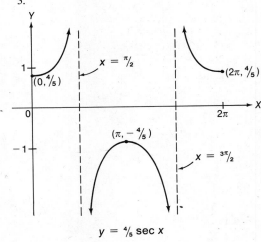

5.

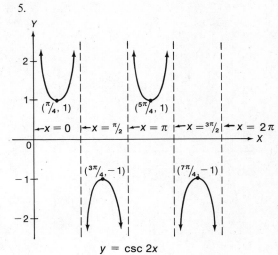

7.

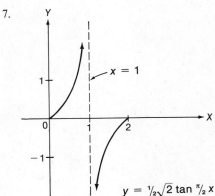

9.

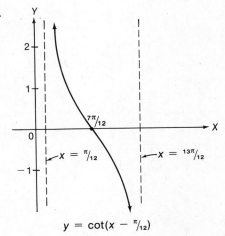

11.

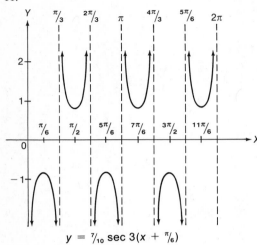

$y = \tfrac{7}{10}\sec 3(x + \tfrac{\pi}{6})$

3.

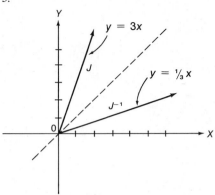

Both J and J^{-1} are functions.
$\mathscr{D}_J = \{x \mid x \geq 0\} = \mathscr{D}_{J^{-1}}$
$\mathscr{R}_J = \{y \mid y \geq 0\} = \mathscr{R}_{J^{-1}}$

13.

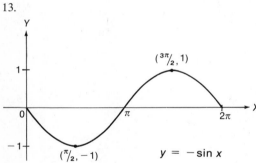

$y = -\sin x$

5.

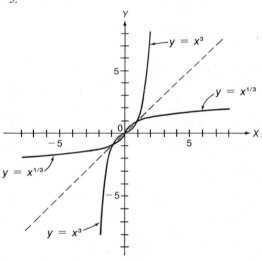

Both L and L^{-1} are functions.
$\mathscr{D}_L = \{x \mid -8 \leq x \leq 8\}$
$\mathscr{R}_L = \{y \mid -2 \leq y \leq 2\}$
$\mathscr{D}_{L^{-1}} = \{x \mid -2 \leq x \leq 2\}$
$\mathscr{R}_{L^{-1}} = \{y \mid -8 \leq y \leq 8\}$

4.4
1.

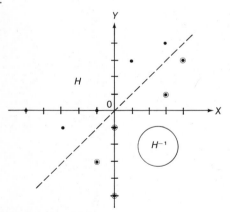

H is a function, H^{-1} is not.
$\mathscr{D}_H = \{-5, -3, -1, 1, 3\} = \mathscr{R}_{H^{-1}}$
$\mathscr{R}_H = \{-1, 0, 3, 4\} = \mathscr{D}_{H^{-1}}$

7.

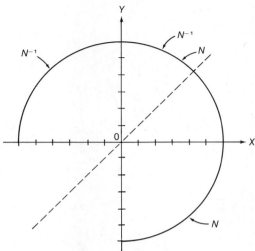

N is not a function, but N^{-1} is a function.
$\mathcal{D}_N = \{x \mid 0 \leq x \leq 6\}$
$\mathcal{R}_N = \{y \mid -6 \leq y \leq 6\}$
$\mathcal{D}_{N^{-1}} = \{x \mid -6 \leq x \leq 6\}$
$\mathcal{R}_{N^{-1}} = \{y \mid 0 \leq y \leq 6\}$

9.

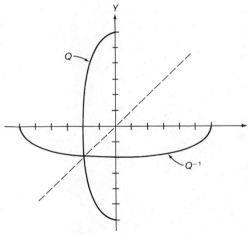

$\mathcal{D}_Q = \{x \mid -2 \leq x \leq 0\}$
$\mathcal{R}_Q = \{y \mid -6 \leq y \leq 6\}$
$\mathcal{D}_{Q^{-1}} = \{x \mid -6 \leq x \leq 6\}$
$\mathcal{R}_{Q^{-1}} = \{y \mid -2 \leq y \leq 0\}$

11. $y = \sqrt{36 - x^2}$

13.
a

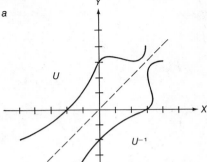

b

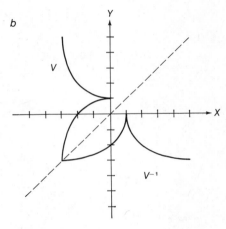

4.5
1. $\pi/6$ 3. $\pi/6$ 5. $-\pi/4$ 7. $5\pi/6$
9. $-\pi/2$ 11. 0.58   13. -0.88
15. $1/\sqrt{3}$ 17. $1/\sqrt{3}$ 19. $1/2$
21. $-\sqrt{3}$ 23. $3/5$ 25. $5/\sqrt{11}$
27. $5/9$ 29. $\sqrt{5/3}$
37. a. 1.18   b. 0.20   c. 1.10

4.6
1.

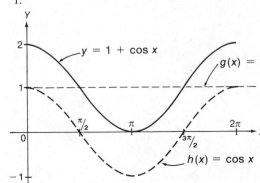

3.

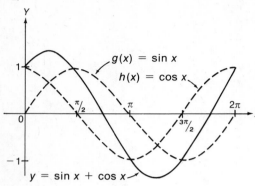

5.

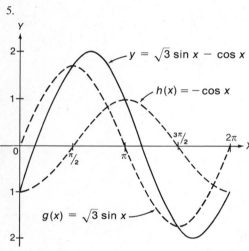

7.

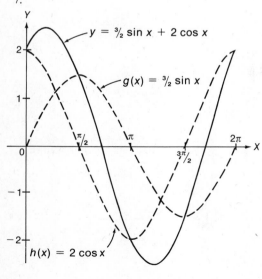

9.

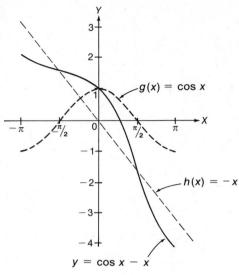

11.

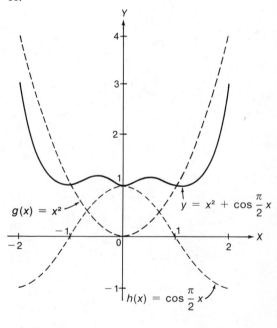

ANSWERS TO SELECTED PROBLEMS

13.

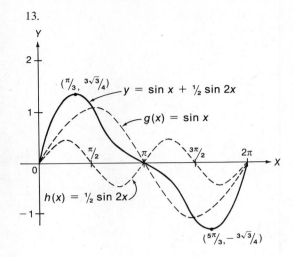

15.

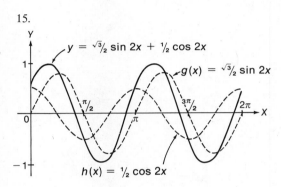

17. a. $y = \sqrt{2} \sin(x + \pi/4)$: Max $(\pi/4, \sqrt{2})$, Min $(5\pi/4, -\sqrt{2})$ b. $y = 2\sin(x - \pi/6)$: Max $(2\pi/3, 2)$, Min $(5\pi/3, -2)$
c. $y = 3/2 \sin[x + \text{Arc}\sin(4/5)]$:
Max $[\pi/2 - \text{Arc}\sin(4/5), 3/2)]$
Min $[3\pi/2 - \text{Arc}\sin(4/5), -3/2)]$
d. $y = \sin 2(x + \pi/12)$: Max $(\pi/6, 1)$ and $(7\pi/6, 1)$, Min $(2\pi/3, -1)$ and $(5\pi/3, -1)$

REVIEW EXERCISES

1. Zeros: $(0, 0), (\pi/3, 0), (2\pi/3, 0)$
   Max $(\pi/6, 1)$, Min $(\pi/2, -1)$
   See the figure at the top of the next column.

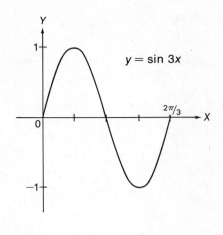

2. Zeros: $(0, 0), (\pi/2, 0)$
   Asymptote: $x = \pi/4$

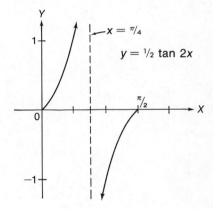

3. Zeros: $(\pi, 0), (3\pi, 0)$
   Max $(0, 1), (4\pi, 1)$, Min $(2\pi, -1)$

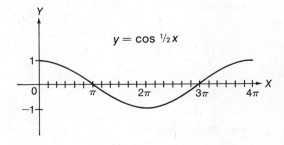

4. Zero: $(2\pi/3, 0)$
   Asymptote: $x = 4\pi/3$
   See the figure at the top of the next page.

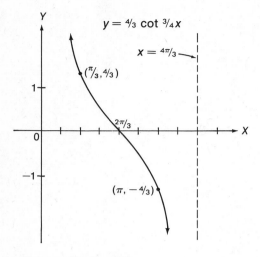

7. Zeros: $(\pi/3, 0), (4\pi/3, 0)$
   Max $(5\pi/6, 1)$
   Min $(11\pi/6, -1)$

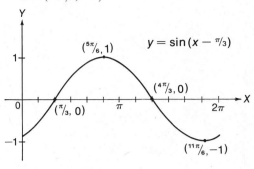

5. Rel. max $(3\pi/2, -4/5)$
   Rel. min $(\pi/2, 4/5)$
   Asymptotes: $x = 0, x = \pi, x = 2\pi$

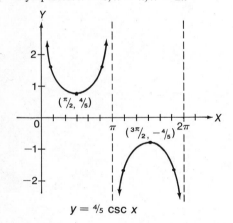

8. Zeros: $(\pi/12, 0), (7\pi/12, 0)$
   Max $(5\pi/6, 3/5)$
   Min $(\pi/3, -3/5)$

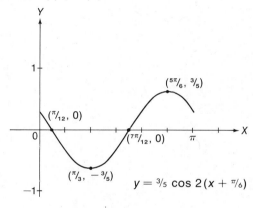

6. Rel. max $(2, -6/5)$
   Rel. min $(0, 6/5), (4, 6/5)$
   Asymptotes: $x = 1, x = 3$

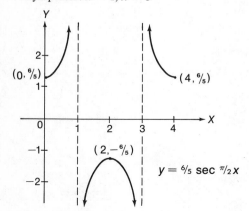

9. Zero: $(1.8, 0)$
   Asymptote: $x = 0.8$

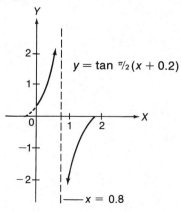

ANSWERS TO SELECTED PROBLEMS 231

10. Rel. max $(7\pi/4, -1)$
    Rel. min $(3\pi/4, 1)$
    Asymptotes: $x = \pi/4, x = 5\pi/4$

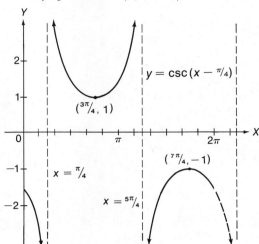

11. $-\pi/6$ 12. $\pi/4$ 13. $5\pi/6$ 14. π
15. $-\pi/4$ 16. $1/3$ 17. $9/5$ 18. $7/24$
19. $2/3$ 20. $7/8$
21. Both A and A^{-1} are functions.
    $\{x \mid 0 \leq x \leq 2\}$
    $\{y \mid -4 \leq y \leq 4\}$

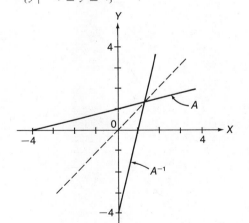

22. Both B and B^{-1} are functions.
    $\{x \mid 0 \leq x \leq \sqrt{15}\}$
    $\{y \mid 1 \leq y \leq 4\}$

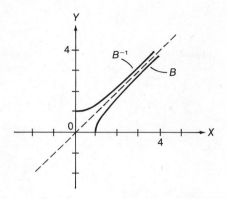

23. C is a function, C^{-1} is not.
    $\{x \mid 0 \leq x \leq 2\}$
    $\{y \mid -4 \leq y \leq 4\}$

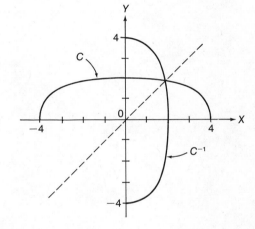

24. Both D and D^{-1} are functions.
    $\{x \mid -1 \leq x \leq 1\}$
    $\{y \mid 0 \leq y \leq \pi\}$

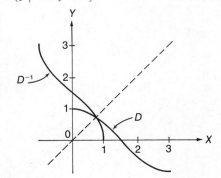

25.

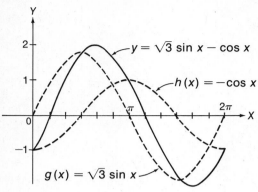

26. See Figure 4.7.26 at the bottom of the page.

27. $y = 2 \sin (x - \pi/6)$
Amplitude: 2
Period: 2π
Phase shift: $\pi/6$

CHAPTER FIVE
5.1
1. $3 + 8i$ 3. $11 + i$ 5. $-3 + 0i$
7. $-1 + i$ 9. $6 + 9i$ 11. $4 - 4i$
13. $-1 + i$ 15. $-5 - 10i$ 17. $7 - 3i$
19. $-5 + 14i$ 21. $25 + 0i$ 23. $29 + 0i$
25. $(a^2 + b^2) + 0i$ 27. $-21 + 20i$
29. $-7 - 24i$ 31. $(x^2 - y^2) + 2xyi$
33. $3 + (5/4)i$ 35. $2 + 3i$
37. $(-3/5) - (6/5)i$ 39. $(5/13) + (12/13)i$
41. $(6/25) - (17/25)i$ 43. $(14/13) - (8/13)i$
45. $0 + i$ 47. $-1 + 0i$ 49. $0 - i$
51. $2 + 11i$ 53. $0 - i$ 55. $-4 - 4i$
57. $2 + i, -2 - i$ 59. $3 + 4i, -3 - 4i$

5.2
1. $1(\cos 300° + i \sin 300°)$
3. $2\sqrt{2}(\cos 135° + i \sin 135°)$
5. $8(\cos 330° + i \sin 330°)$
7. $5(\cos 216°50' + i \sin 216°50')$
9. $-5 + 0i$ 11. $-3 + 3\sqrt{3}i$
13. $-4 - 4i$ 15. $-3.83 - 3.21i$
17. $1 + 2i, -1 + 0i, -1 + i$
19. $5 - i, -1 + 3i, 8 - i$
21. $20(\cos 250° + i \sin 250°)$,
$5(\cos 150° + i \sin 150°)$,
$(1/5)(\cos 210° + i \sin 210°)$,
$100(\cos 40° + i \sin 40°)$
23. $54(\cos 120° + i \sin 120°)$,
$(3/2)(\cos 200° + i \sin 200°)$,
$(2/3)(\cos 160° + i \sin 160°)$,
$81(\cos 320° + i \sin 320°)$
25. *a.* 6 cis 90°   *b.* 10 cis 170°   *c.* 2 cis 345°
d. 6 cis 315°
27. *a.* 80 cis 240°   *b.* (4/5) cis 240°
c. (5/4) cis 120°   *d.* 64 cis 120°
29. *a.* 24 cis 240°   *b.* (3/2) cis 60°
c. (2/3) cis 300°   *d.* 36 cis 300°
31. $1 - 2i, -1 + 2i$

5.3
1. 81 cis 200°   3. 8 cis 300°   5. 4 cis 180°
7. 64 cis 180°   9. cis 240°
11. 2 cis 20°, 2 cis 140°, 2 cis 260°
13. $\sqrt{2}$ cis 0°, $\sqrt{2}$ cis 60°, $\sqrt{2}$ cis 120°, $\sqrt{2}$ cis 180°,
$\sqrt{2}$ cis 240°, $\sqrt{2}$ cis 300°   15. cis 45°, cis 225°
17. $1, i, -1, -i$
19. $\frac{1}{2} + \frac{1}{2}\sqrt{3}i, -1, \frac{1}{2} - \frac{1}{2}\sqrt{3}i$
21. $1.93 + 5.18i, -1.41 + 1.41i, -.518 - 1.93i$
23. $2 - 3i, -2 + 3i$ 25. $0, 0, 0$

5.4
1. $(3, 90°)$ 3. $(2\sqrt{2}, 315°)$ 5. $(5, 126°50')$
7. $(0, -4)$ 9. $(-3, 3\sqrt{3})$
11. $(-9.40, -3.42)$ 13. $(2\sqrt{3}, -2)$
15. $\rho = \sin \theta$ 17. $\rho = 4 \sec \theta$
19. $2/(1 - \sin \theta), -2/(1 + \sin \theta)$
21. $x^2 + y^2 = 9$ 23. $x^2 + y^2 - 2x = 0$
25. $y^2 - 4x - 4 = 0$

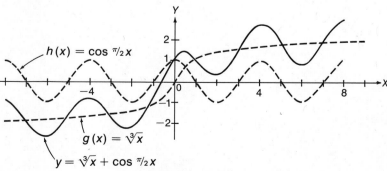

FIGURE 4.7.26

ANSWERS TO SELECTED PROBLEMS   233

27.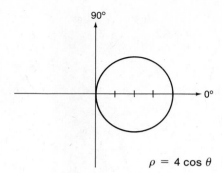
$\rho = 4 \cos \theta$

29.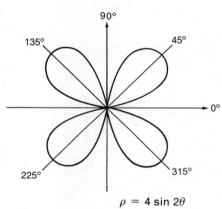
$\rho = 4 \sin 2\theta$

31.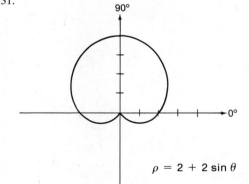
$\rho = 2 + 2 \sin \theta$

33.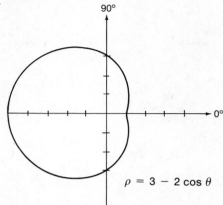
$\rho = 3 - 2 \cos \theta$

35.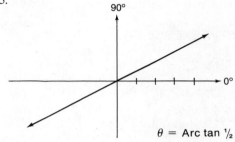
$\theta = \text{Arc tan } \tfrac{1}{2}$

37.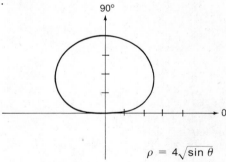
$\rho = 4\sqrt{\sin \theta}$

39.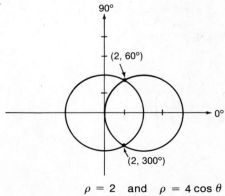
$\rho = 2 \quad \text{and} \quad \rho = 4 \cos \theta$

41.

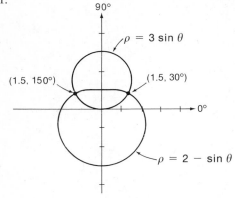

33.

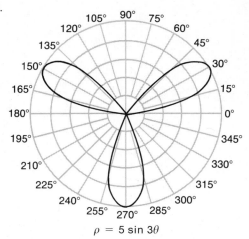

$\rho = 5 \sin 3\theta$

REVIEW EXERCISES
1. $2 + i$ 2. $1 + 2i$ 3. $0 + 5i$
4. $-1 + 2i$ 5. $17 - 6i$ 6. $-13 - 18i$
7. $3 - 4i$ 8. $7 - 24i$ 9. $9/10 - (7/10)i$
10. $1/13 + (18/13)i$ 11. $0 - i$ 12. $0 + i$
13. $-9 + 46i$ 14. $-11 + 2i$
15. $-4 + 0i$ 16. $4 - 4i$ 17. $1 + i$
18. $1/16 + (1/16)i$ 19. $4 - i$ and $-4 + i$
20. $3 - 4i$ and $-3 + 4i$ 21. $-3\sqrt{3} + 3i$
22. $3 - 3i$ 23. $-2\sqrt{2} - 2\sqrt{2}i$
24. $-1 - \sqrt{3}i$
25. a. $48(\cos 120° + i \sin 120°)$
 b. $3(\cos 240° + i \sin 240°)$
 c. $144(\cos 0° + i \sin 0°)$
 d. $16(\cos 240° + i \sin 240°)$
26. a. $18(\cos 340° + i \sin 340°)$
 b. $2(\cos 240° + i \sin 240°)$
 c. $36(\cos 220° + i \sin 220°)$
 d. $9(\cos 100° + i \sin 100°)$
27. 2 cis 50°, 2 cis 122°, 2 cis 194°, 2 cis 266°, 2 cis 338°
28. 2 cis 30°, 2 cis 150°, 2 cis 270°
29. $3/(1 + \cos \theta)$ or $-3/(1 - \cos \theta)$
30. $\rho = -4 \cos \theta = 0$
31. $4x^2 + 3y^2 + 8y - 16 = 0$
32. $y^2 = 1 - 2x$

34.

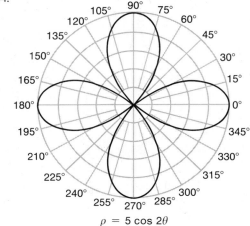

$\rho = 5 \cos 2\theta$

35.

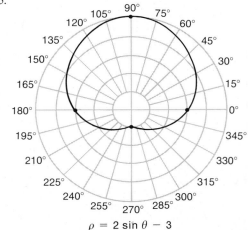

$\rho = 2 \sin \theta - 3$

36.

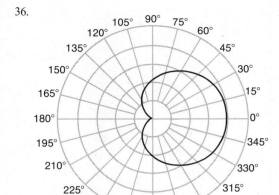

$\rho = 2(\cos\theta + 1)$

37.

$\rho = \dfrac{2}{1 - \cos\theta}$

38.

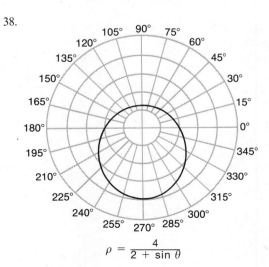

$\rho = \dfrac{4}{2 + \sin\theta}$

CHAPTER SIX
6.1

1. $a = \dfrac{200 \sin 35°}{\sin 73°} \approx 120$, $b = \dfrac{200 \sin 72°}{\sin 73°} \approx 199$,
$\gamma = 73°$, $\mathscr{A} = 100\, a \sin 72° \approx 11{,}400$ sq units

3. $\beta = 42°00'$, $a = \dfrac{270 \sin 33°10'}{\sin 42°00'} \approx 221$,
$c = \dfrac{270 \sin 104°50'}{\sin 42°00'} \approx 390$
$\mathscr{A} = 135 a \sin 104°50' \approx 28800$ sq units

5. $\alpha = \text{Arc sin}\left(\dfrac{3}{4}\sin 40°\right) \approx 28°50'$,
$\gamma = 140° - \alpha \approx 111°10'$, $c = \dfrac{40 \sin \gamma}{\sin 40°} \approx 58.0$
$\mathscr{A} = 600 \sin \gamma \approx 560$ sq units

7. $\alpha = \text{Arc sin}\, (5.23/7.07) \approx 47°40'$,
$\beta = 90° - \alpha \approx 42°40'$, $b = 7.07 \sin \beta \approx 4.76$,
$\mathscr{A} = \tfrac{1}{2}(5.23)b \approx 12.4$

9. $\dfrac{5000 \sin 25°}{\sin 120°} \approx 2440$ yd

11. $b = \dfrac{10 \sin \beta}{\sin 8°20'} \approx 56.4$ mi,
$\beta = 171°40' - \gamma \approx 125°10'$,
$\gamma = \text{Arc sin}\,(5 \sin 8°20') \approx 46°30'$

6.2

1. One solution.
$\beta = \text{Arc sin}\left(\dfrac{2}{3}\sin 36°\right) \approx 23°00'$,
$\gamma = 144° - \beta \approx 121°$, $c = \dfrac{12 \sin \gamma}{\sin 36°} \approx 17.5$

3. Two solutions.
$\gamma = \text{Arc sin}\left(\dfrac{4}{3}\sin 20°20'\right) \approx 27°40'$,
$\alpha = 159°40' - \gamma \approx 132°00'$, $a = \dfrac{150 \sin \alpha}{\sin 20°20'} \approx 321$
$\gamma' = 180° - \gamma \approx 152°20'$, $\alpha' = 159°40' - \gamma' \approx 7°20'$,
$a' = \dfrac{150 \sin \alpha'}{\sin 20°20'} \approx 55.1$

5. One solution. $\gamma = 90°$, $\beta = 65°10'$,
$b = \dfrac{63 \sin 65°10'}{\sin 24°50'} \approx 136$

7. One solution.
$\beta = \text{Arc sin}\,(0.1 \sin 135°20') \approx 4°00'$,
$\gamma = 44°40' - \beta$, $c = \dfrac{35.2 \sin \gamma}{\sin 135°20'} \approx 32.7$

9. No solution.

11. Two solutions.
$\alpha = \text{Arc sin}\left(\dfrac{512}{475}\sin 65°30'\right) \approx 78°50'$,

$\gamma = 114°30' - \alpha \approx 35°40'$, $c = \dfrac{475 \sin \gamma}{\sin 65° 30'} \approx 304$

$\alpha' = 180° - \alpha \approx 101°10'$,
$\gamma' = 114°30' - \alpha' \approx 13°20'$
$c' = \dfrac{475 \sin \gamma'}{\sin 65°30'} \approx 120$

13. 273 mi or 127 mi 15. 856 ft

6.3

1. $b = \sqrt{164 - 80\sqrt{3}} \approx 5.04$
3. $a = \sqrt{784} = 28$
5. $c = \sqrt{1525 - 1500 \cos 50°} \approx 23.7$
7. $\alpha = 90°$
9. $\gamma = \text{Arc cos } (269/360) \approx 41°40'$
11. $\beta = \text{Arc cos } (-4.01/25.6) \approx 99°00'$
13. $c = \sqrt{1649} \approx 40.6$
15. $\cos \beta = (a^2 + c^2 - b^2)/2ac$
17. $b = \sqrt{544 - 480 \cos 82°30'} \approx 21.9$,
$\gamma = \text{Arc sin } (20 \sin 82°30'/b) \approx 64°50'$,
$\alpha = 97°30' - \gamma \approx 32°40'$
19. $c = \sqrt{16.49 + 16 \cos 35°10'} \approx 5.44$,
$\alpha = \text{Arc sin } (2.5 \sin 35°10'/c) \approx 15°20'$,
$\beta = 35°10' - \alpha \approx 19°50'$
21. $\gamma = \text{Arc cos } (5.77/17.98) \approx 71°20'$,
$\beta = \text{Arc cos } (11.05/20.3) \approx 57°00'$,
$\alpha = 180° - (\beta + \gamma) \approx 51°40'$
23. $\gamma = \text{Arc cos } (-295/484) \approx 127°30'$,
$\beta = \text{Arc cos } (179/220) \approx 35°30'$,
$\alpha = 180° - \beta - \gamma \approx 17°00'$
25. $\sqrt{7925 + 7700 \cos 35°} \approx 119$ mi
27. $d = \sqrt{3321 - 3240 \cos 80°} \approx 52.5$ mi,
course $= 50° - \text{Arc sin } (36 \sin 80°/d) \approx 7°30'$
29. $\mathscr{A} = 525 \sin \gamma \approx 517$ sq units,
$\gamma = \text{Arc cos } (361/2100)$.
31. *a.* acute *b.* right *c.* obtuse
    d. acute *e.* obtuse *f.* acute

6.4

1. $|\bar{v}_x| = 10 \cos 75° \approx 2.59$,
$|\bar{v}_y| = 10 \sin 75° \approx 9.66$

3. $|\bar{v}_x| = 5.6 \cos 35° \approx 4.59$,
$|\bar{v}_y| = 5.6 \sin 35° \approx 3.21$
5. $|\bar{v}| = \sqrt{65} \approx 8.06$, $\theta = \text{Arc tan } (4/7) \approx 29°40'$
7. $|\bar{v}| = \sqrt{164} \approx 12.8$,
$\theta = \text{Arc sin } (-.8) \approx -29°40'$
9. $|\bar{v}| = \sqrt{72.32 - 71.68 \cos 80°} \approx 7.74$,
$\theta = \alpha - 20° \approx 34°30'$,
$\alpha = \text{Arc sin } (6.4 \sin 80°/|\bar{v}|) \approx 54°30'$
11. $\sqrt{369 + 360 \cos 20°} \approx 26.6$
13. Airspeed: $\sqrt{10625 - 5000 \cos 80°} \approx 98.8$ mph,
Heading: $320° + \omega \approx 334°30'$,
$\omega = \text{Arc sin } (25 \sin 80°/|\overline{H_a}|) \approx 14°30'$
15. $|\overline{W}| = \sqrt{37925 - 37700 \cos 8°} \approx 24.3$ mph,
from $293° + \alpha \approx 341°10'$, $\alpha = \text{Arc sin } (130 \sin 8°/|\overline{W}|) \approx 48°10'$

6.5

1. $8.96''$,
$h = \dfrac{\omega \cos \alpha \cos \beta}{\sin(\alpha + \beta)} - r\left[\csc\left(\dfrac{\alpha + \beta}{2}\right)\cos\left(\dfrac{\beta - \alpha}{2}\right) - 1\right]$
3. Hint: TN bisects angle RTU, since RT and UT are tangent to the circle; hence $\gamma = \dfrac{\alpha + \beta}{2}$.

Angle $\theta = \gamma - \alpha = \dfrac{\alpha + \beta}{2} - \alpha = \dfrac{\beta - \alpha}{2}$.

5. 164 rpm 7. *a.* 0.9854ℓ *b.* 0.9848ℓ
c. 0.9854ℓ *d.* 0.9903ℓ

9. $y = \tan \theta_0 x - \dfrac{16}{v_0^2 \cos^2 \theta_0} x^2$;
quadratic equation; parabola
11. *a.* $\sqrt{2}/8 \approx 0.177$ sec *b.* $\sqrt{2}/4 \approx 0.354$ sec
13. $1.25 \sin 25° \approx 0.528$ sec
15. $50 \sin 50° \approx 38.3$ ft; $2.5 \sin 25° \approx 1.06$ sec
17. See Figure 6.5.17 at bottom of page
19. *a.* 25 *b.* $5\sqrt{15} \approx 19.4$ *c.* 22.4 *d.* 25

REVIEW EXERCISES

1. $a = 6$, $\alpha = 82°50'$, $\beta = 55°50'$, $\gamma = 138°40'$
2. $\alpha = 22°40'$, $\gamma = 117°20'$, $c = 13.8$
3. No solution.

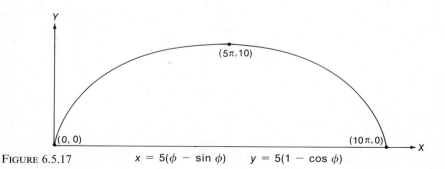

FIGURE 6.5.17 $x = 5(\phi - \sin \phi)$ $y = 5(1 - \cos \phi)$

ANSWERS TO SELECTED PROBLEMS

4. $\alpha = 78°30', \beta = 44°30', \gamma = 57°00'$
5. $b = 5.09, \alpha = 20°40', \gamma = 34°20'$
6. $\beta = 56°00', \gamma = 87°10', c = 82.0,$
$\beta' = 124°00', \gamma' = 19°10', c' = 26.9$
7. 70.9 sq units   8. 367 sq units
9. 34.0 sq units
10. Hint: $\sin(180° - x) = \sin x$
11. $|\bar{v}_x| = 13.4, |\bar{v}_y| = 14.9$ 12. $112°20'$
13. 114 mph; $7°40'$; $307°40'$
14. a. $\dfrac{2860 \sin 20°}{32} \approx 30.6$ sec
b. $\left(\dfrac{2860 \sin 20°}{8}\right)^2 \approx 15{,}000$ ft
c. $\dfrac{2860^2 \sin 40°}{32} \approx 164{,}000$ ft
d. $\dfrac{2860 \sin 20°}{16} \approx 61.1$ sec
15. 31.3; $9°20'$
16. a. 95.4 mi   b. $077°$ c. 57.2 minutes

APPENDIX A
1. a. 0.3797   b. 0.7619   c. 0.6184
d. 1.790   e. 1.294   f. 0.4962   g. 0.1142
h. 0.7541
3. a. $29°44'$ b. $42°27'$ c. $39°21'$
d. $1°50'$ e. $59°35'$ f. $66°25'$ g. $47°42'$
h. $74°13'$
5. a. 0.5261   b. 0.5097   c. 0.0170

d. 0.6435
7. a. 0.927   b. 0.114   c. 1.161   d. 0.183

APPENDIX B
1. 0.1736   3. 0.9989   5. 0.1118

APPENDIX C
1. a. 0.7300   b. 0.5185   c. 0.9031
d. 2.7846   e. 4.1959   f. $9.7160 - 10$
g. $6.4771 - 10$ h. $7.4698 - 10$
3. a. 3.97   b. 7.59   c. 184   d. 5770
e. 0.0392   f. 88.6   g. 0.167   h. 0.00593
5. a. $9.7973 - 10$ b. 0.2804
c. $9.8035 - 10$ d. 0.3583
7. a. $23°20'$ b. $46°10'$ c. $71°30'$
d. $66°50'$
9. a. 1.97   b. 7.64   c. 153   d. 0.244
11. a. 42.2   b. 116   c. 8360   d. 2.52
e. 36.3
13. a. $56°00'$ b. $71°10'$ c. $46°40'$
d. $22°30'$
15. a. $B = 36°20', b = 3.74, c = 6.32$
b. $B = 60°40', a = 18.2, c = 37.2$
c. $A = 42°00', B = 48°00', a = 883$
d. $A = 37°50', B = 52°10', c = 0.579$
17. a. $\gamma = 47°10', b = 110, c = 82.8$
b. $\alpha = 40°50', \gamma = 79°20', c = 759$
c. $\alpha = 15°20', \beta = 152°30', b = 624, \alpha' = 164°40',$
$\beta' = 3°10', b' = 74.4$

Index

Absolute value of a complex number, 127
Accuracy, 7, 148
Addition identities, 65–71
Addition of ordinates, 114–118
Algebra, review of, 21–27
Altitude of a triangle, 18
Ambiguous case, 154–158
Amplitude of complex number, 127
Amplitude of periodic function, 90
Angles, 14–16
    acute, 15
    alternate-interior, 16
    central, of a regular polygon, 19
    corresponding, 16
    coterminal, 29
    of depression, 8
    of elevation, 8
    initial side of, 29
    least positive, 32
    negative, 29
    obtuse, 15
    positive, 29
    quadrantal, 32, 41
    reference, 46
    right, 15
    of rotation, 77
    sides of, 14, 29
    standard position, 28
    straight, 15
    supplementary, 15
    terminal side of, 29
    vertex of, 14
    vertical, 15

Angular velocity, 177
Antilogarithm, 190
Arc cos, Arc sin, Arc tan, 109–113
Arc length, 29
Area
    of a circle, 20
    of a circular sector, 34
    of a circular segment, 34
    of a regular polygon, 10
    of a triangle, 12, 18, 146–149
Asymptotes, 100

Bearings, 163

Characteristic of a logarithm, 189
Circle, 19–20
Cofunctions, 3, 5
Complementary angles, 3, 5
Complex numbers, 121–138
    absolute value of, 127
    conjugate of, 123, 129
    equality of, 122, 129
    graphical representation of, 127
    powers of, 123, 132–138
    product of, 122, 130
    quotient of, 123, 130
    roots of, 125, 132–138
    sum of, 122
    trigonometric form of, 127
Components of a vector, 8, 168
Cosecant function, 36
    graph of, 56
Cosine of an acute angle, 3

Cosine function, 36
    graph of, 90
Cosines, Law of, 159–165
Cotangent of an acute angle, 3
Cotangent function, 36
    graph of, 100
Coterminal angles, 29
Counterexample, 72
Cycle, 90

Degree measure, 15
De Moivre's theorem, 133
Direction angle of a vector, 167
Distance formula, 25
Domain
    of a function, 51, 106
    of a relation, 105
Double-angle formulas, 73

Equations
    conditional trigonometric, 80–84
    in polar coordinate form, 140–143
    quadratic, 23–25

Functions, 51
    domain of, 51, 106
    even, 52
    inverse of, 104–108
    inverse trigonometric, 109
    odd, 52
    periodic, 53
    range of, 51, 106
    trigonometric, 51–56

Geometry, review of, 14–20
Graphs
    of a complex number, 127
    of inverse functions and relations, 104–108
    of inverse trigonometric functions, 109–113
    of trigonometric functions, 51–56, 87–104

Half-angle formulas, 54
Hipparchus, 1–2
Hypotenuse of right triangle, 2

Identities, 59–78
    addition and subtraction, 65–71
    double angle, 73–78
    fundamental, 59, 60
    half angle, 74–78
    Pythagorean, 59, 73
    quotient, 59
    reciprocal, 59

Imaginary number, 122
Inclination of a line, 69
Initial side of an angle, 29
Integers, 22
Interpolation, 180–184
Inverse
    of a relation, 104
    of trigonometric functions, 109–113

Law of Cosines, 159–165
Law of Sines, 149–151, 153–158, 161
    ambiguous case, 154–158
Line segment, 14
Logarithms
    common, 188–192
        table of, 208–209
Logarithms
    of trigonometric functions, 188–192
        table of, 210–217

Magnitude of a vector, 167
Mantissa of a logarithm, 189
Minute, 15

Numbers
    imaginary, 122
    irrational, 23
    natural, 22
    rational, 22
    real, 22

Ordered pair of numbers, 23, 51, 104
Ordinates, addition of, 114–118

Perimeter of a polygon, 10, 19
Period of a function, 53
Phase shift, 98
Polar coordinates, 138–143
Polygon, 18
    area of, 10
    regular, 18
Pythagorean relation, 4, 17

Quadrants, 31
Quadratic equation, 23
Quadratic formula, 24

Radian measure, 29–32
Range
    of a function, 51, 106
    of a relation, 105
Rational numbers, 22
Ray, 14
Real numbers, 22, 23

INDEX

Reciprocal relations, 37, 38, 59
Reference angle, 46
Relation, 104
Resultant of vectors, 167
Right angles, 15
Right triangles, 2–9, 16–17
    trigonometric ratios of, 2–3

Secant function, 36
    graph of, 100
Second, 15
Sector of a circle, 19
    area of, 34
Segment of a circle
    area of, 34
Set builder notation, 23
Sets of numbers, 22, 23
Side adjacent, 2
Side of an angle, 14
Side opposite, 2
Sine of an acute angle, 3
Sine function, 36
    graph of, 53–55, 88–91
Sines, Law of, 149–151, 153–158, 161
Slope of a line, 26, 70

Slope-intercept form, 27
Standard notation for triangles, 3
Standard position of an angle, 28
Summation formulas, 72, 73

Tangent of an acute angle, 3
Tangent function, 36
    graph of, 55, 100
Terminal side of an angle, 29
Transversal, 16
Triangles, 16–18
    area of, 12, 18, 146–149
    congruent, 18
    equilateral, 17
    isosceles, 17
    right, 2–9, 16–17
    similar, 5, 18
Trigonometric equations, 80–84
Trigonometric form of a complex number, 127
Trigonometric identities, 59–78

Vector, 8
Velocity, angular, 177
Vertex of an angle, 14